Essentials
of Safety

third edition

Essentials of Safety

Alton L. Thygerson
Brigham Young University

Prentice Hall, Inc., Englewood Cliffs, New Jersey 07632

Library of Congress Cataloging-in-Publication Data

Thygerson, Alton L.
  Essentials of safety.

  Rev. ed. of: Safety. 2nd ed. ©1976.
  Includes bibliographies and index.
  1. Accidents—United States—Prevention.
  2. Safety education—United States. I. Thygerson,
  Alton L. Safety. II. Title.
  HV675.T427 1986 363.1 85-16770
  ISBN 0-13-287863-1

Editorial/production supervision and
  interior design: Barbara DeVries
Cover design: Joseph Curcio
Manufacturing buyer: Harry Baisley

The second edition of this book was published under the title
Safety: Concepts and Instruction.

Printed in the United States of America

10 9 8 7 6 5 4 3 2 1

ISBN 0-13-287863-1 01

Prentice-Hall International (UK) Limited, *London*
Prentice-Hall of Australia Pty. Limited, *Sydney*
Prentice-Hall Canada Inc., *Toronto*
Prentice-Hall Hispanoamericana, S.A., *Mexico*
Prentice-Hall of India Private Limited, *New Delhi*
Prentice-Hall of Japan, Inc., *Tokyo*
Prentice-Hall of Southeast Asia Pte. Ltd., *Singapore*
Editora Prentice-Hall do Brasil, Ltda., *Rio de Janeiro*
Whitehall Books Limited, *Wellington, New Zealand*

Dedicated to the most important people in my life — my family

Contents

3

USING ACCIDENT DATA   20

4

THE SAFETY MOVEMENT   29

5

HAZARDS, RISKS, AND RISK-TAKING   39

Preface

Essentials of Safety is a clear, straightforward presentation of the foundations upon which safety is based.

Although there are many professional and lay opinions concerning the accident problem, the foundations for a science of accident prevention and injury control remain inadequate. This book is an attempt to present the needed foundation.

The book deals with terms such as "accident" and "safety" like no other book does. Other relevant content includes a chapter on risk-taking, another on determining causes of accidents, and a third devoted to strategies for accident prevention and injury control.

Featured in this book are anecdotal "boxes" which provide pertinent world records, accounts of unusual accidents, historical background, and relevant quotes.

Indebtedness to numerous persons is freely acknowledged. In some cases it has been possible to document this fact in specific citations and acknowledgments. The helpful cooperation of numerous publishers in giving permission to quote from their publications is gratefully acknowledged. Appreciation is extended to those at Prentice-Hall who helped in the development and production of this book.

Alton L. Thygerson

1

Introduction

The safety of the American people has never been better. In this century we have witnessed a remarkable reduction in life-threatening accidents along with other public health problems (e.g., infectious and communicable diseases). Nevertheless, accidents rank as the most frequent cause of death for persons between the ages of one and thirty-eight. Moreover, trauma (accidental death and injury) is a seriously neglected public health problem in the United States.

It is the thesis of this book that further improvements in the safety of the American people can be achieved through a commitment to accident prevention and control.

Since 1912, the accidental death rate in the United States has been reduced from eighty-two per 100,000 persons to forty (a reduction of over 50 percent). The reduction in the overall rate during a period when the nation's population more than doubled has resulted in 2,400,000 fewer accidental deaths than would have been the case if the rate had not been reduced.[1]

Much of the credit for this must go to earlier efforts of prevention, based on new knowledge obtained through research. But some of the recent gains are due to measures by government to eliminate hazards and reduce their undesirable effects in the event of an accident. However, we have the scientific knowledge to decrease even further the trauma toll from accidents in which the loss of human life is great.

The message should be clear that much of today's accidental death and disability can be averted. Modest expenditures of dollars can yield high dividends in terms of both lives saved and an improvement in the quality of life.

The case for preventing accidents and controlling their effects has been established in part by a number of accomplishments. Some of the most successful examples are packaging poisonous agents in child resistant containers, equipping refrigerator doors to open easily from the inside, and banning thousands of unsafe toys.

Compared with many diseases of far less consequence, the prevention of death, injuries, and property damage due to accidents has received relatively little scientific attention. Moreover, the expenditure of money for research in the safety area is quite small when compared with the money spent on other public health problems (such as heart disease, cancer, etc.).

Although individual behavior is clearly important to accident causation, emphasis on personal responsibility ignores the important role of the social, political, economic, and physical environments that largely determine behavior. Often pleas for safer lifestyles are likely to be ineffective. Efforts to modify individual behavior will continue, but other successful approaches should be utilized that give automatic protection without requiring any special action on the part of these being protected.

WHAT IS AN ACCIDENT?

The definition of an "accident" is complicated and not easily understood. Though all of us have had an accident and assume that we know what an accident is, it is more difficult to define than you may think.

Read through each of the statements in Table 1-1 and place a check mark in the blank space if you believe that the statement represents an accident as the term is used within the field of safety.

There are two major attributes of accidents—unintended "causes" and undesirable "effects." These two factors seem to constitute the major elements of most definitions of the term "accident." For example, in the United States the most widely accepted definition of an accident is the one used by the National Safety Council: "that occurrence in a sequence of events which usually produces unintended injury, death, or property damage."[2] Two subfactors essential in the definition of an accident are: (1) the suddenness of the event (in seconds or minutes) and (2) the damage resulting from one of the forms of physical energy (e.g., mechanical, thermal, chemical, etc.).

Unintended Causes

What is meant by an "unintended cause"? The so-called "act of God," such as being hit by lightning or drowning in a flood, is generally considered an unintended cause. At the other end of the scale is an "act of man," where it is clear that the cause was due to some human behavior.

Most people experience daily unintended events that are called accidents (e.g.,

TABLE 1-1. Which Are Accidents?

Directions: Check the items which you consider to be an accident.

_____ 1. A depressed woman died from an overdose of Valium.

_____ 2. An animal trainer was bitten to death by a tiger.

_____ 3. A woman fell off a trolley car with no apparent injuries but then sued on the claim that she became a nymphomaniac because of the fall.

_____ 4. While hurrying to meet a manuscript deadline, a secretary made a typing mistake.

_____ 5. A college student became very angered when a professor forgot an appointment.

_____ 6. Two people were trampled to death by irate soccer fans rioting over a referee's decision.

_____ 7. A professional golfer received minor injuries after being struck by lightning.

_____ 8. A driver was angered when another vehicle barely missed him.

_____ 9. A college student caught a cold from his girl friend.

_____ 10. During a bank robbery, a security guard was shot and killed.

_____ 11. Residents of St. George, Utah, show high cancer death rates after receiving excessive amounts of radiation fallout from nuclear bomb testing in Nevada during the 1950s.

_____ 12. The family car's right fender was bent while a teenager family member was backing the car out of the garage.

_____ 13. Due to twenty-five years of jarring while operating a bulldozer, the operator developed a lumbar (low back) disc pain.

_____ 14. A two-year-old child died after swallowing aspirin obtained from his mother's purse.

_____ 15. An elderly couple died from hypothermia after being stranded in their car during a winter blizzard.

These statements should have been checked: 2, 7, 12, 14, and 15. These situations meet the criteria for an accident as defined in the field of safety.

mistakes in typing, forgetting an appointment) but are not physically injurious. Usually only those events resulting in physical injury, death, and/or property damage are labeled as accidents in the field of safety.

BOX 1-1

Whatever can go wrong, will.
—Murphy's Law

It is important to differentiate between events that are intentional and those that are unintentional.[3] Unintentional events include accidents (drownings, falls, etc.) and disasters (tornadoes, earthquakes, etc.). Intentional acts include crime (homicide, theft, assault, arson, etc.), suicide, wars, and riots. Therefore, if intent can be shown to precede the injury, the occurrence becomes a criminal act, not an accident. The focus of the field of safety is upon unintentional causes which have resulted in undesirable effects (death, injury, and/or property damage).

It is also important to note that the term "unintended" covers only damage from one of the forms of physical energy in the environment (e.g., mechanical, thermal, chemical, electrical, or ionizing radiation). Thus, if a person inadvertently swallows poison and is damaged, the event is called an accident. However, if that same person swallows a polio virus and is injured, the result is called a disease and is rarely considered accidental.

Edward Suchman[4] indicates that the term "accident" is more likely to be used when the event manifests the following three major characteristics:

1. Degree of expectedness—the less the event could have been anticipated, the more likely it is to be labeled an accident.
2. Degree of avoidability—the less the event could have been avoided, the more likely it is to be labeled an accident.
3. Degree of intention—the less the event was the result of a deliberate action, the more likely it is to be labeled an accident.

These characteristics could include many daily activities such as losing things or forgetting appointments, but within the context of the safety field, physical injury, death, and/or property damage must result. In other words, an injury indicates that an accident happened. The injury also occurs over a relatively short period of time (seconds or minutes).

Undesirable Effects

What is meant by "undesirable effects"? Physical injury in the medical sense can range from minor cuts to death. At any rate, the term "injury" usually refers to damage resulting from sudden exposure to physical or chemical agents, while the result of long-term exposure is usually classified as disease.

Death, injury, and/or property damage—undesirable effects of an unintentional event—do *not* in themselves constitute the accident, but are the results of it. In other words, injury, death, and/or property damage only indicate that an accident has occurred if they result from a sudden or short-term exposure to physical and/or chemical agents.

BOX 1-2

> Whatever can happen to one man can happen to every man.
>
> —*Seneca*

For example, if a driver falls asleep at the wheel but awakes in time to avoid a collision, the event involves no physical injury or damage and is not considered an accident.

An injury could be psychological in nature. Examples include anger, fear, and embarrassment. Such result would not ordinarily be classified as an accident.

BOX 1-3 *UNUSUAL BUT TRUE*

> Bob Aubrey of Ottawa, Ontario had been blind for eight years, tripped over his guide dog and banged his head on the floor. His sight was instantly restored.
>
> —National Safety Council, *Family Safety*

All consequences of accidents need not be entirely negative or undesirable. For example, though a child is burned by touching a hot pan or pot, he learns very quickly that pans and pots can be hot and can cause pain. Thus, this child avoids future contact with hot pans and pots (or possibly even any hot object) and is not burned again. This child's experience, though temporarily painful, has a long-term positive effect. See Table 1-2.

BOX 1-4 *UNUSUAL BUT TRUE*

> Edwin Robinson, legally blind and partially deaf, was standing under a poplar tree near his Falmouth, Maine, home during an electrical storm. A bolt of lightning struck him, restoring his sight and hearing.
>
> —National Safety Council, *Family Safety*

Another example is the after effects of a disaster. Though human life may be taken, survivors learn that living in certain river bottoms in the spring of the year or being near a beach during a hurricane or tsunami should be avoided.

It is not suggested that everyone try to experience things in order to determine consequences. Rather, all people should gain from others' experiences. This is the basis for all instruction and education—a shortcut to experience, but also an avoidance of hazards which may have undesirable effects.

TABLE 1-2. Accidental Scientific Discoveries

DISCOVERY	DISCOVERER	YEAR
Electrical current	Luigi Galvani of Italy	1781
Practical photography	Louis Daguerre of France	1837
Vulcanized rubber	Charles Goodyear of the United States	1839
X-rays	Wilhelm Roentgen of Germany	1895
Radioactivity	Henri Becquerel of France	1896
"Popsicle" (trade name)	Frank Epperson of the United States	1923
"Teflon" (trade name)	Roy J. Plunkett	1938

Some safety experts question whether injuries or damages should be included in a definition of the term "accident" since a resulting injury is the outcome and does not constitute the accident. Why a person is injured is a separate question from why he was involved in an accident.

The two attributes found in most definitions of an accident (unintentional causes and undesirable effects) are independent of each other. It is, for example, possible for a person to fall down the stairs (the unintended event) without hurting himself (the injury). The two factors can assist in determining the appropriate countermeasures to be used in combating the accident problem. "Accident prevention" pertains to the efforts (such as education, prohibitions, licensing) to deal with the unintentional causes of accidents. "Injury control" pertains to the mitigation (rescue, safety belts, emergency medical care, helmets) of injury, death, and/or property damage.

"Accident prevention" has been criticized on the basis that it is not effective enough and does not convey the idea of the basic problem (injuries) and of the desired end (reducing losses due to injuries). Some experts show evidence that countermeasures to the injury problem should not be limited to prevention, but should include any stage of the injury-producing process. The term suggested is "injury control" rather than "accident prevention." this may appear to be an improvement, but it makes no distinction between accidental (unintentional) injury and intentional injury. It would appear that property damage is not clearly identified either, since "injury" usually refers in most cases only to physical human body damage.

Preventing accidents and their undesirable effects requires a clear concept of what an accident is. Unfortunately, even safety experts disagree on its definition. In fact, the many definitions of an accident complicate accident prevention and in-injury control. Moreover, the common definition and fatalistic connotation of the word "accident" appear to be an obstacle since the word implies, to many people, that something unexpected and unpleasant has occurred, but it:

- couldn't be helped, it was an accident.
- was inevitable and would have happened to anyone.
- was unforeseen, and therefore, uncontrollable.
- is not our fault, and therefore we shouldn't be blamed.

Nevertheless, the word is too deeply entrenched in the language to go out of use. In summary, the word "accident," as commonly used in the field of safety, refers to a sudden unintended event which results in death, injury, and/or property damage from one of the forms of physical energy.

WHAT IS SAFETY?

There seems to be no general acceptance of what is implied or denoted in the term "safety." Kenneth Licht[5] has identified five distinct ways in which the word "safety" is used:

I. Examples of Safety or Its Derivative Used in a Health Context
   A. Visitors to exotic lands are cautioned that the public drinking supply may not be *safe*.
   B. Air pollution was so bad last winter that it was *unsafe* for elderly people and those with respiratory problems to go outside.
   C. Many persons still consider it *unsafe* to store canned foods in their original containers.
   D. Even persons with sensitive skin may use this product with complete *safety*.

II. Examples of Safety or Its Derivative Used in a "Security Context
   A. Always carry your credit cards in a *safe* place.
   B. We guarantee the highest interest rates with complete *safety*.
   C. Our recently expanded campus police force has made this the *safest* campus in the state.
   D. Ladies, for your personal *safety*, learn self-defense techniques at our health club.

III. Examples of Safety or Its Derivative Used in a "Special" or "Technical" Context
   A. Don't be half-*safe;* use . . . . deodorant.
   B. The *safety* we scored in the final quarter beat State and made our homecoming a complete success.
   C. Be sure the *safety* is on until you're ready to fire.
   D. It is *safe* to assume that most people prefer comfort to distress.

IV. Examples of Safety or Its Derivative Used in an "Accident Prevention" Context
   A. It's a good practice to have your car *safety*-checked before your vacation trip.
   B. The right way is the *safe* way.
   C. Worn tires are more *unsafe* at today's expressway speeds than they were when average speeds were lower.
   D. Platform shoes may be stylish, but they're certainly *unsafe*.

V. Examples of Safety or Its Derivative Used in an "Accident Mitigation" Context (Note: This category deals with efforts to reduce (mitigate) the consequences of accidents. It takes up where accident prevention leaves off. Hard hats, for example, have little to do with accident prevention, but everything to do with accident mitigation. A closer examination of the concept of accidents and accident mitigation will follow.)
   A. *Safety* rules require hard hats be worn on this job.
   B. More than half the states now have *safety* regulations requiring eye-protection articles be worn in all shops and labs.
   C. Energy-absorbing steering columns, padded dashes and corner posts, recessed handles, knobs, and other hostile projections have made the new cars much *safer*.
   D. A first-aid course is essential to anyone interested in a career in *safety*.

In his book, *Of Acceptable Risk*, William Lowrance[6] explores the meaning of safety. A thing is safe, according to Lowrance, if its risks are judged to be acceptable; and judging the acceptability of that risk (judging safety) is a matter of personal and social value judgment. Safety can change from time to time and be judged differently in different contexts.

Lowrance points out that the term "safety" is vague as used in the past and has long been misused. He contends that risks, not safety, are measured. A thing is safe if its attendant risks are judged to be acceptable.

Lowrance differentiates by saying that:

Measuring Risk—a measure of the probability of harm—described as a scientific activity.

Measuring Safety—a judgment of the acceptability of risks—described as a value-judging activity.

We all face varying degrees of risks. An example of an "acceptable risk" would be an act almost every one of us performs many times a week: opening our mail. Have you ever gotten a paper cut while tearing open the flap of an envelope? That is an injury associated with a very common item, a paper envelope. Yet we do not consider a paper envelope hazardous. The risk of injury involved in tearing open an envelope is acceptable.

Objects with sharp edges present an interesting example of "unacceptable risks" which also can be considered "acceptable" depending on the circumstance. Anything with a sharp edge is hazardous. But we simply accept the risk whenever we use a knife, and we use it carefully to minimize the hazard. However, very young children do not understand these concepts of *risk* and *hazard*, and they certainly cannot exercise the same kind of care as adults. So while a sharp edge on a kitchen knife constitutes an acceptable risk to an adult, the risk becomes unacceptable for children. Similarly, a sharp edge on a child's toy presents an unacceptable risk.

Remember, Lowrance defines *risk* as something we can *measure* and *safety* as something we must *judge*. Take the matter of the paper envelope. We can measure the severity of the cuts and the probability or the frequency with which they occur. This gives us a measure of the *risk* involved in opening a paper envelope. Next, we judge whether this risk is worth taking. Yes, we say, it is. We judge that paper envelopes are *safe*.

In our example of the knives, we can measure the risk of a cut. We judge that knives are safe for adults, who can handle them carefully, but are unsafe for children, who cannot. On a child's toy, a sharp edge produces such a high probability of harm, poses such a high risk, that we judge it unsafe. See page 43 for more information on how to judge what is safe.

WHY STUDY ABOUT ACCIDENTS AND SAFETY?

Awareness

Individuals should be aware of the main accident problems. Have you ever visited a place and then been surprised at how often you heard or read about that place afterward? The place had been mentioned just as frequently before you visited it, but you never noticed these comments until you were acquainted with it; thereafter, you found meaning in each reference because it related to something that had become familiar. Similarly, if we are aware of a particular accident problem, we notice each reference to it in the newspaper, possibly take time to read a magazine article about it, or become more attentive when it comes up in conversation. In this way we constantly increase our knowledge of a problem and the validity of our judgments about it.

Factual Knowledge

An intelligent analysis must rest upon facts. It makes little sense to discuss an accident problem unless someone in the group knows what he or she is talking about. Although fact gathering will not solve any problem automatically, it is impossible to analyze a problem until the facts have been collected, organized, and interpreted.

Misconceptions about safety, accidents, and injuries exist because all the facts may not be known or presented. Unfortunately, these misconceptions and myths persist yet, in spite of educational efforts. (See Figures 1-1 and 1-2.) As a self-check for misconceptions, see how well you perform on the test in Table 1-3.

FIGURE 1-1
Does lightning ever strike twice in the same place? Photo courtesy of the *Deseret News.*

FIGURE 1-2
Myths exist about the rattlesnake. The venom of bees, wasps, and hornets causes more deaths in the United States each year than are caused by the venom of rattlesnakes. Photo courtesy of the Utah Division of Natural Resources.

The Science of Accident Prevention and Injury Control

It is valuable that we have a general understanding of how and why accidents occur, how people are affected by accidents, and how we can deal with them. This provides a frame of reference within which we can catalog data and study specific problems. If we have a thorough understanding of the science of accident prevention and injury control we can intelligently organize and analyze data on any particular accident. In addition, this understanding enables us to interpret new data

TABLE 1-3   Test on Safety Misconceptions

Directions: Place a check mark in front of the statement if you believe it to be true.

_____	1. A person hit by lightning usually dies instantly.
_____	2. Lightning never strikes twice in the same place.
_____	3. Red is the hunter's best clothing color.
_____	4. A rattlesnake gives warning before striking.
_____	5. It is impossible to stay afloat in water for long with clothes on.
_____	6. If a boat overturns, you should swim to shore.
_____	7. A drowning person always comes up for air three times.
_____	8. Coffee will help sober up a drunk.
_____	9. Smaller vehicles can stop in less time and distance than larger ones.
_____	10. Pumping the brakes helps stop a car more quickly on icy roads.
_____	11. A rattlesnake has to be coiled up to strike.
_____	12. Sharks won't attack close to shore.
_____	13. Moss always grows on the north side of a tree trunk.
_____	14. There is a difference between "flammable" and "inflammable."
_____	15. The only way you can get poison ivy is to touch the plant.

Number 10 should have been checked as true. All the other statements are false.

correctly and to keep up to date. Accident data often become obsolete and accident problems may change considerably within a few years. For example, there is a change in concern about motorcycle and snowmobile death and injury statistics and a decrease in concern about abandoned refrigerators and ultrathin plastic clothing covers as potential hazards. Yet, if the individual understands the science of accident prevention, he or she will not find it hard to interpret new data and understand new accident trends.

Our attitudes and values determine the meanings we find in the facts we observe. A study of some widespread fallacious attitudes toward accidents may help show why people react to facts so differently. Such a study may also help show why we may always have accident problems.

Listed below are some fallacious beliefs presented by authors Richardson, Hein, and Farnsworth:

1. The "other fellow" concept, whereby it is assumed that accidents happen to other people but won't happen to you.
2. The "your number's up" concept, whereby it is assumed that "when your number is up," you will get hurt and there is nothing you can do about it.
3. The "law of averages" concept, whereby accidents and injuries are shrugged off as due to inevitable statistical laws.
4. The "price of progress" concept, whereby accidents are rationalized as the inevitable price of scientific advancement.
5. The "spirit of '76" concept, whereby living dangerously is glorified and safety measures are regarded as sissy.
6. The "act of God" concept, whereby accidents are seen as divinely caused—for punishment or for some purpose unknown to us.[7]

A Sense of Perspective

Some people find the study of accidents upsetting. Just as people are frightened by all the diseases they find listed in a medical textbook, some individuals are disturbed by the great amount of expressed or implied criticism of our society that they find in a safety course. For others, awareness of imperfections which result in accidents may become an obsession. They are so distressed with the tragedy, suffering, and waste caused by an accident-plagued society that they fail to see more encouraging aspects of the total picture. We need a sense of perspective if we are to see without exaggeration or distortion.

Appreciation of the Proper Role of the Expert

Opinions are not equally valuable. If we want a useful opinion on why our head throbs or our car stalls, we ask the appropriate experts. However, as we ponder accidents, we hesitate to ask the safety expert and confidently announce our own opinions, perhaps after discussing the questions with others who know no more than we.

This contradiction stems from our failure to distinguish between *questions of knowledge* and *questions of value*. In questions of knowledge there are right and wrong answers, whereas in questions of value there are differences of opinion. The layman and the expert are equally qualified to answer questions of value, but they are not equally qualified to answer questions of knowledge. For instance, the question of whether leisure time should be used in viewing operas or football games is a matter of value. But the question of whether a four-phase driver education program is more or less effective than a two-phase driver education program requires expert knowledge to answer.

Stated in simple terms, *the role of the expert is not to tell people what they should want, but to tell them how they may best get what they want.* Even experts are not infallible; all may be wrong on a given issue. When experts disagree, no one answer should be considered positive or final. Experts in safety agree that accidents are caused and do not "just happen"; the layman who feels an accident was an event which could not have been avoided reveals his ignorance.

In the field of safety, the function of the expert is to provide accurate descriptions and analyses of accidents and to show laymen what consequences may follow each countermeasure proposal. However, experts have met with limited success when they have attempted to tell people what would be best for them in a given situation. It usually takes a near-accident, an accident, or even a tragedy before people become sufficiently concerned and motivated to take overt action. Recall from page 8 that experts measure risk, while safety is unusually determined by personal and social value judgments.

The task of the individual is to learn how to recognize an expert and guide his own thinking by expert knowledge rather than by guesswork.

NOTES

1. National Safety Council, *Accident Facts* (Chicago: National Safety Council, 1984), p. 10.
2. ___., p. 97.
3. Kenneth F. Licht, "Safety: What Is It?" *School Safety World Newsletter*, 1, no. 3 (Summer 1973), 3.
4. Edward A. Suchman, "A Conceptual Analysis of the Accident Phenomenon," in William Haddon, Jr., Edward A. Suchman, and David Klein, eds., *Accident Research* (New York: Harper & Row, Publishers, Inc., 1964), p. 276.
5. Kenneth F. Licht, "Safety and Accidents—A Brief Conceptual Analysis and a Point of View," *Journal of School Health*, XLV, no. 9 (November 1975), 530-31.
6. William W. Lowrance, *Of Acceptable Risk* (Los Altos, Calif.: William Kaufmann, Inc., 1976), p. 8.
7. Charles E. Richardson, Fred V. Hein, and Dana L. Farnsworth, *Living* (Glenview, Ill.: Scott, Foresman & Company, 1975), pp. 339-40.

2

Magnitude of the
Accident Problem

EXTENT OF ACCIDENT OCCURRENCE

Reading statistics can be quite dull. Reading accident statistics may be repulsive as well. However, accident statistics of any kind tell us only where and how accidents happened, but not why. Let us consider some general accident statistics.

Every year in this country approximately one hundred thousand accidental deaths are reported by the National Safety Council. This averages out to one death every five minutes.[1] About nine million disabling injuries are reported annually. Disabling injuries are not reported on a nationwide basis, therefore the total numbers of injuries are estimates and should not be compared from year to year. Death accounts are more accurate than data on nonfatal injuries, which are often under-reported and under-estimated; for many types of accidents records are almost nonexistent.

Table 2-1 ranks accidents with other major causes of death. The tragedy of the high accidental death loss is that trauma kills thousands who otherwise could expect to live long and productive lives, whereas those afflicted with heart disease, cancer, stroke, and many chronic diseases usually die late in life.

The human suffering and financial loss from preventable accidental death constitute a public health problem second only to the ravages of ancient plagues or world wars. In making comparisons of fatal accidents with war, it must be kept in

TABLE 2-1   Accidents and Other Causes of Death

RANK	CAUSE OF DEATH
1	Heart disease
2	Cancer
3	Stroke (cerebrovascular disease)
4	Accidents
	Motor vehicle
	Falls
	Drowning
	Fires, burns
	Suffocation — ingested object
	Poisoning by solids and liquids
	Firearms
	Poisoning by gases and vapors
5	Pneumonia
6	Diabetes mellitus
7	Chronic liver disease, cirrhosis
8	Atherosclerosis
9	Suicide
10	Homicide
11	Certain conditions originating in perinatal period
12	Nephritis and nephrosis

Source: National Safety Council, *Accident Facts*, 1984, p. 8.

mind that nearly everyone is exposed to accidents, but relatively few are exposed to war deaths. The point is that we are very much concerned about war and its effects, but to be unconcerned about a domestic problem which causes as much or even more damage betrays a paradox in human thought and values.

One in four Americans is annually injured badly enough to require medical attention or activity restriction for at least one day. Accidents cause more deaths each year than all infectious diseases combined. Although accidental deaths were reduced by over 50 percent during the past seventy-five years, trauma from accidents still poses a major problem.

The accident picture in the United States is grim; yet it is fair to assume that without organized safety efforts and safety education, America's accident record would be even more shocking than it is. Following heart disease, cancer, and stroke, accidents are the fourth principal cause of death in the United States. Accidents are the leading cause of death among those persons aged one to 38 years.[2]

We often think of accidents only when they are catastrophes because these events make newspaper headlines. From a statistical point of view, a catastrophe is an accident in which five or more lives are lost.[3] It is significant that a very small percentage of accident catastrophes occur as a result of natural forces such as floods, hurricanes, tornadoes, and earthquakes (see Figure 2-1). Rather, the majority of newspaper headline catastrophes are caused by some kind of human failure which results in airplane crashes, mine cave-ins, explosions, and the like. However, the

FIGURE 2-1 Disasters occur in all parts of the world. That they will come is certain. Photo courtesy of the American National Red Cross.

record of natural catastrophes reveals that relatively few lives are lost in this way when contrasted with the total number of deaths resulting from other types of unspectacular, unpublished accidents.

SURVEYS OF ACCIDENTS AMONG THE GENERAL PUBLIC

The National Health Survey, conducted by the National Center for Health Statistics, is a survey of households to collect information on the number of injuries sustained by household members during the two weeks prior to the survey interview. (It should be noted that other health information is also surveyed.) The total number of injuries is then estimated for the entire United States based upon these findings from approximately 40,000 households. We realize that differences in definition produce different injury totals when we compare the results of the National Health Survey with the totals presented by the National Safety Council.

An examination of accident injury data gathered from the general population suggests the following conclusions:

1. Estimates of the number of accidents based solely on the data of deaths or reported injuries are incomplete.

2. Data derived from these sources point out the need for a standard accident definition and reporting system.

The question of definition—and *who does the defining*—is a crucial one. For example, one may never detect or report the housewife who burns her fingers. This type of incident occurs commonly in daily living and illustrates how large numbers of accidents remain unknown.

ESTIMATES OF THE COST OF ACCIDENTS

Reliable estimates of the overall cost per annum of accidents in the United States are difficult to make, and those who have investigated the problem conclude that even approximate figures would be inaccurate.

We can compare accident costs to an iceberg—only a small portion that we can actually see and measure appears above water. The indirect and hidden costs form the rest of the iceberg, the part below water that is not easily measured. Examples of costs are listed in Table 2-2.

TABLE 2-2 Some Social and Economic Consequences of Accidents

	SOCIAL		ECONOMIC
1	Grief over the loss of loved ones	1	Costs of rescue equipment required
2	Loss of public confidence	2	Accident investigation and reporting
3	Loss of prestige	3	Fees for legal actions
4	Deterioration of morale	4	Time of personnel involved in rescue
5	Denial of education	5	Medical fees (doctors, hospitals, etc.)
6	Lack of guidance for children	6	Disability costs of personnel badly injured
7	Psychological effects of a change in standard of living	7	Replacement cost of property damaged or lost
8	Psychic damages affecting behavior	8	Slowdown in operations while accident causes are determined and corrective actions taken
9	Embarrassment	9	Loss of income
10	Lost pride	10	Loss of earning capability
11	Inconvenience	11	Rehabilitation costs for those who have lost limbs, mental abilities, or physical skills
12	Adversely affected interpersonal relationships (anger, resentment, etc.)	12	Funeral expenses for those killed
		13	Pensions for injured persons or for dependents of those killed
		14	Training costs and lower output of replacements
		15	Production loss for employer

GEOGRAPHICAL DISTRIBUTION OF ACCIDENTS
IN THE UNITED STATES

Regional Accident Rates

Rates of reported accidents vary widely from one region to another. In accidental death rates, the Middle Atlantic region ranks lowest. The Rocky Mountain and Southern States regions have the highest rates of accidental deaths. Various explanations could be offered for these wide variations, but they would probably still not account for all the variations found.

Rural-Urban Variations

Accident death rates tend consistently to decrease with greater density of population. The difference in rates between the most densely populated and the least densely populated communities is sizable.

GENERAL CHARACTERISTICS OF THE ACCIDENT VICTIM

The "typical" accident victim exists only as a vague abstraction created from statistical figures. Nevertheless, statistics enable us to say:

1. Most accident victims are males. After the first year of life, males have more accidents than females, at all age levels.
2. Most accident victims are young. Accidents are the leading cause of death for persons aged one to thirty-eight years; however, the rate for accidental deaths is highest for those over seventy years of age, with the fifteen to twenty-four year age group next highest.
3. Accidental death rates are highest in rural areas. The Rocky Mountain and Southern regions of the United States have the highest accident rates.
4. Most accidental deaths occur in a motor vehicle, but most injuries occur in the home. This is explained by the high speeds involved in motor-vehicle accidents resulting in death; whereas, the home is not as lethal, but because of time spent there accounts for the most injuries.
5. Most accidents happen to the victim in a cyclical manner, reaching peaks in frequency on certain days and at a particular time of the year. Weekends and the summer months of June, July, and August have the highest rates.

As will be discussed in Chapter 3, accurately assessing facts presents a real problem.

SOCIAL AND ECONOMIC CONSEQUENCES OF ACCIDENTS

Accidents produce consequences of grave importance in terms of death, injury, and property damage. Table 2-2 presents in brief form some social and economic implications of accidents.

TABLE 2-3   Years of Potential Life Lost by Causes of Death

CAUSE OF DEATH	YEARS OF POTENTIAL LIFE LOST BEFORE AGE 65 BY PERSONS DYING IN 1982	ESTIMATED NUMBER OF PHYSICIAN CONTACTS DECEMBER 1981
ALL CAUSES (includes data not shown separately)	9,429,000	84,586,000
Accidents	2,367,000	4,610,000
Malignant neoplasms	1,809,000	1,403,000
Diseases of heart	1,566,000	4,956,000
Suicides, homicides	1,314,000	—
Cerebrovascular diseases	256,000	557,000
Chronic liver diseases and cirrhosis	252,000	86,000
Pneumonia and influenza	118,000	1,067,000
Chronic obstructive pulmonary diseases and allied conditions	114,000	2,025,000
Diabetes mellitus	106,000	2,312,000

Source: Centers for Disease Controls.

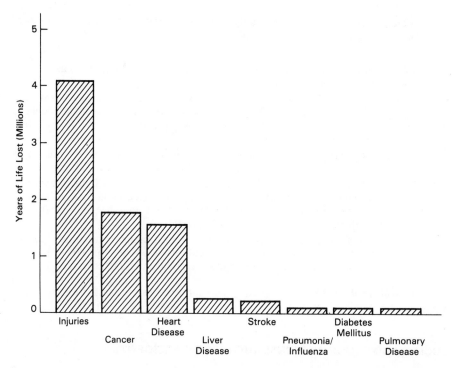

FIGURE 2-2 Potential Years of Life Lost Prior to Age 70 from Eight Leading Causes of Death. From *The Injury Fact Book* by Susan P. Baker, Brian O'Neill, and Ronald S. Karpf (Lexington, Mass.: Lexington Books, D.C. Heath and Company, Copyright 1984, D.C. Heath and Company).

We should consider the serious consequences of accidents in terms of their disruptive effects on the home, on the family, and on employment. The notions of "life years lost" and "working years lost" are helpful in gauging the magnitude of the accident problem. Table 2-3 and Figure 2-2 show us that accidental deaths are responsible for a greater number of years of potential life lost than are deaths from any other cause.[4]

NOTES

1. National Safety Council, *Accident Facts*, 1983, p. 11.
2. All statistical data in this book are from the National Safety Council, *Accident Facts*, 1984, unless otherwise noted.
3. Metropolitan Life Insurance Company, *Statistical Bulletin* (January 1975), p. 6.
4. Nelson S. Hartunian, Charles N. Smart, and Mark S. Thompson, *The Incidence and Economic Costs of Major Health Impairments* (Lexington, Mass.: D.C. Heath and Company, 1982).

3

Using Accident Data

The first decennial census was taken in the United States in 1790. However, it was not until after the turn of the twentieth century that accident data were collected systematically.

ACCIDENT FACTS

Issued annually by the National Safety Council, *Accident Facts* now constitutes a most reliable source of nationwide frequency for all types of accidents reported in the United States.[1] In addition to presenting annual reports on the four principal classes of accidents, *Accident Facts* presents data on rates and trends; hence this compendium is an available source for evaluating the relative efficiency of accident-prevention attempts. The National Safety Council has been aided in the collection and interpretation of accident data by many organizations, companies, and individuals.

OTHER SOURCES OF ACCIDENT STATISTICS

Except for statistical studies made by individual investigators which are usually limited to small samples of accident victims, the large-scale gathering of statistical data is necessarily a task of the government. At present, there are several sources available for accident statistics:

National Center for Vital Statistics. This government agency summarizes death certificates. Details are brief. Reports often state "victim fell," or "victim poisoned"—and little else.

National Center for Health Statistics. This federal agency conducts the National Health Survey annually. This survey of about forty thousand households is primarily concerned with health problems. Accidents are included, but are a relatively small portion of the survey. Its weakness is that it does not get enough information. Although forty thousand households are contacted annually, the interviewer asks for accident experience of household members during only the previous two-week period. This totals about eight thousand person-years of exposure annually. Thus, if a type of accident occurs only twenty-five thousand times a year, we can expect to find only one case in the National Health Survey for that year. From the survey, then, one gets overall totals—a broad perspective of what is going on; but one does not get the details needed. Numerical differences between National Health Survey totals and the National Safety Council totals occur and are due mainly to differences in the injury definitions used.

National Electronic Injury Surveillance System (NEISS). This is a national data system used to determine the magnitude and scope of consumer product-related injuries. NEISS gathers data daily through a computer system located throughout the United States. Certain accident cases are assigned for follow-up interviews to collect detailed information. The details, available through on-site visits so that photographs can be taken, the product examined or collected, and measurements made or done by telephone interviews, establish the relationship between the victim, the consumer product, and the environment.
  NEISS enables the identification of hazards. NEISS data are supplemented by death certificate records for fatalities throughout the United States and by newspaper accounts.

National Center for Statistics and Analysis. This center is under the jurisdiction of the National Highway Traffic Safety Administration and operates data collection and analysis programs to support motor vehicle and highway safety efforts.
  The Center's major data collection systems include:

Fatal Accident Reporting System (FARS) Which is a census of every fatal motor vehicle accident in the United States, containing information obtained from state records.
National Accident Sampling System (NASS) which is a nationally representative sample of all police-reported traffic accidents and contains detailed information beyond what is available in state record systems. Data are collected by trained accident investigatiors under contract to NHTSA.

State Agencies. Organizations such as state departments of vital statistics, highway departments, and safety councils frequently collect, analyze, and distribute information on the incidence of accidents in their particular geographical area and special interest and responsibility.

Research Reports. Excellent sources of information on specific accident problems are available through the safety literature. (See Table 4-1.) Many of the research reports are located in public and university libraries.

NEED FOR ACCIDENT DATA

Why do we need accident data? Such data provide evidence that an accident and injury problem exists, and that it is important to know of its existence. Actually, severe accidents are rare events in the life of most individuals and few persons can get any clear understanding of their importance as a social and economic problem solely from their own experience. A second reason for having accident data is the need to evaluate the effectiveness of accident prevention and injury control countermeasures.

Data usually comes in statistical form, and statistics are of two types:

1. *Descriptive statistics* provide methods of organizing, summarizing, and communicating data. Most of the accident and injury statistics are descriptive in nature. Total sums, ratios, averages, and percentages are examples of descriptive statistics.

2. *Inferential statistics* provide methods for making inferences from the descriptive data. Research reports usually use inferential statistics. Examples of such statistics include statistics indicating differences between groups and statistics showing relationships. A college course in statistics is needed to properly understand and interpret the use of inferential statistics.

PROBLEMS OF STATISTICAL ANALYSIS

Statistics play an important role in accident prevention. The reasons statistics are not used more is probably because of what some call a "statistical neurosis."

An early statistician coined a phrase that has since been repeated to generations: "It's not that the figures lie—it's that the liars figure." Although deceptive figures do appear in accident statistics, it is probable that the largest number of statistical studies are compiled by those desiring to inform rather than misinform. Nevertheless, a statistic that misleads through honest error can be just as confusing as one that is deliberately constructed to misrepresent. In the following paragraphs we will discuss several sources of error found in accident statistics; these may be divided into the following groups: sources of error in *collection, presentation,* and *interpretation.*

Sources of Error in the Collection
of Accident Statistics

Deliberate suppression. There are several forms of suppression. One can be attributed to administrative self-protection rather than to darker motives. Keeping the accident rate down is a perennial concern of administrators. Frequently, a slightly injured worker is immediately returned to the job to keep the factory record spotless, even if the injured worker should be home after the accident resting from the shock and caring for his injury.

Only a few accident statistic errors in collection result from distortions introduced deliberately or negligently. The following sources of error are highly difficult or impossible to eradicate.

Failure to complain. Since an accident is not known until someone reports it, any statistic that purports to represent the total number of any type of accident that has taken place is necessarily an informed guess based on the addition of an estimated number of *unreported* accidents of the same category. Certain types of accidents tend to be reported with much greater accuracy than others. For example, deaths resulting from motor-vehicle accidents are usually reported quite accurately because there are legal requirements to make such reports and great value is placed upon human life. However, motor-vehicle accidents causing less than one hundred dollars worth of damage may go unnoticed and even unreported as a result of legal stipulations and one hundred dollar deductible insurance coverage which may not require the reporting of small accidents. Even reports of damage costs may be totally in error since most estimated costs are quite subjective in nature.

Geographical variations in the definition of certain accidents. There is one obstacle hindering a uniform system of accident reporting which will not be overcome until all jurisdictions employ similar definitions for classifying accidents. For example, different definitions yield different statistics. In Belgium and Italy a person is considered a highway fatality only if he dies at the accident scene. In France an interval between injury and death that would still permit such a classification is three days, twenty-four hours in Spain, thirty days in Britain, and a year in the United States and Canada.

Differences in the reporting of accidents within the same area. An administrator has his biases. If a relative or a person with high status has an accident, the administrator may never report it in order to keep a record clear of any blemish that might hinder future promotions and advancements for that person. Another example was given by William E. Tarrants at a National Safety Congress:

A mid-western manufacturing plant with a fairly stable injury frequency rate decided to hire a full-time safety engineer to see if this rate could be cut down. Within a short time after the safety engineer was hired, the injury frequency rate nearly doubled. Should we conclude that the safety engineer caused these accidents and quickly fire him? A closer look revealed that the safety engineer instituted a new accident investigation and reporting system which produced more reports of disabling injuries, thus increasing the frequency rate.[2]

Lack of uniformity in collecting and recording techniques. Tallying the number of known accidents is a considerably more complicated process than it would seem to be. A motor-vehicle accident is usually witnessed, whereas there may be no witnesses to a home accident involving minor cuts or burns, making accurate reporting more difficult.

Sources of Error in the Presentation of Accident Statistics

When we consider statistical data there is probably no assertion more misleading than the frequently heard statement: "The figures speak for themselves." Because long columns of figures convey an impression of factuality, it is essential to discuss the more common misunderstandings that arise from faulty presentation of data.

Misleading use of simple sums rather than rates. The fact that Town *A* records fifty motor vehicle deaths as compared with one hundred reported by Town *B* does not necessarily mean that Town *B* is twice as accident-ridden as Town *A*. If Town *B* has four times the population of Town *A* the reverse is true. Total accident figures do not become meaningful until they are transformed into *rates* (ratios or percentages) based on the total population under consideration. Similarly, this principle applies to changes in the incidence of accidents. Town *C*, in 1990, may report twice as many accidents as it did in 1970—yet its population may have doubled during this period. If this is true, then the accident rate remains the same.

Misleading use of averages and percentages. The precaution of translating simple sums into averages and percentages does not assure the proper presentation of data; these measures can be misleading as the raw totals. Consider the following statement: "The average number of accidents per family in the Lakeview residential area is three per year." This statement gives the impression that each family living in the Lakeview residential area has about the same number of accidents each year. This assumption may not be correct at all. A count reveals that five families live in the area. Three households report one accident, whereas the fourth family reports two accidents that year. The fifth family has a total of ten among its members. Thus the "average" figure obtained by adding the total number of accidents and dividing by five is mathematically accurate but misleading.

As the illustration demonstrates, an *average* is meaningless without information about the variation of the measures that compose it.

Pitfalls of graphic presentations. For some readers, long columns of figures are not only impressive but downright intimidating. For this reason statisticians and publicists often present their findings graphically or pictorially. Although this method presents material in quick, easy-to-comprehend form, it can also mislead the reader.

Sources of Error in the Interpretation of Accident Statistics

The "self-evident" conclusion. Mark Twain observed that there are three kinds of lies—plain lies, damned lies, and statistics. This humorous statement implies that statistics can be manipulated to support any point of view. The use of statistics, however, should not be completely dismissed, for statistics mislead and confuse only when one does not know how to interpret them. There are elaborate formulas for determining the significance and reliability of a statistic. This text offers some simple tests which the reader can apply.

There are some basic criteria which should be applied in analyzing or interpreting accident information and data:

1. Who compiled and reported the information (e.g., background, reputation, and experience of the individual or group)?
2. Why is the information provided (e.g., to inform the public, sell a product, or other motives)?
3. When was the information compiled (e.g., recently or long ago)?
4. How is the information presented?
5. Where did the information come from (e.g., research findings or personal opinion)?
6. Is the information accurate and reliable (e.g., sound conclusions or questionable data)?
7. What of it (e.g., important or insignificant)?

The problems with many of the sources of accident data are: (1) information incomplete, (2) reporting on only severe types, (3) information not easily available or not made public, (4) inaccuracies, and (5) variations in definition of accident and/or injury.

The statement that "four times more fatalities occur on the highways at 7 P.M. than at 7 A.M." can be misleading. People fail to realize that more people are killed in the evening than in the morning simply because more people are on the highways at that hour to be killed. An amusing example of nonsense statistics is found in Table 3-1.

TABLE 3-1 Pickles Will Kill You

Pickles will kill you: Every pickle you eat brings you nearer to death. Amazingly the "thinking man" has failed to grasp the terrifying significance of the term "in a pickle." Although leading horticulturists have long known that Cucumis sativus possesses an indehiscent pepo, the pickle industry continues to expand.

Pickles are associated with all the major diseases of the body. Eating them breeds wars and communism. They can be related to most airline tragedies. Auto accidents are caused by pickles. There exists a positive relationship between crime waves and consumption of this fruit of the cucurbit family. For example:

Nearly all sick people have eaten pickles. The effects are obviously cumulative.

99.9% of all people who die from cancer have eaten pickles.

100% of all soldiers have eaten pickles.

96.8% of all communist sympathizers have eaten pickles.

99.9% of all the people involved in air and auto accidents ate pickles within 14 days preceding the accident.

93.1% of juvenile delinquents come from homes where pickles are served frequently.

Evidence points to the long-term effects of pickle eating:

Of the people born in 1839 who later dined on pickles, there has been a 100% mortality.

All pickle-eaters born between 1849 and 1859 have wrinkled skin, have lost most of their teeth, have brittle bones and failing eyesight—if the ills of eating pickles haven't already caused their death.

Even more convincing is the report of a noted team of medical experts: rats force-fed with 20 pounds of pickles per day for 30 days developed bulging abdomens. Their appetites for WHOLESOME FOOD were destroyed.

In spite of all the evidence, pickle-growers and packers continue to spread their evil. More than 120,000 acres of fertile soil are devoted to growing pickles. Our per capita yearly consumption is nearly four pounds.

Eat orchid petal soup. Practically no one has as many problems from eating orchid petal soup as they do from eating pickles.

The field of safety and accident prevention is replete with figures, statistics, numbers, and ratios. However, there is little literature available regarding the proper use of accident statistics. William Tarrants, at a National Safety Congress, presented information which he felt was pertinent:

> During World War II about 375 thousand people were killed in the United States by accidents and about 408 thousand were killed in the armed forces. From these figures, it has been argued that it was not much more dangerous to be overseas in the armed forces than to be at home. A more meaningful comparison, however, would consider rates of the same age groups. This comparison would reflect adversely on the safety of the armed forces during the war—in fact, the armed forces death rate (about 12 per thousand men per year) was 15 to 20 times as high, per person per year, as the over-all civilian death rate from accidents (about 0.7 per thousand per year).
>
> The same fallacy is noted in the widely publicized statistical appraisal which states that "Sometime during the Korean War we passed a grim milestone. One day an American soldier fell in battle. He was the millionth American soldier to die in our wars since the nation was born. A few months later the 1,000,000th American perished in a modern highway traffic accident. [*Author's note:* The two millionth traffic fatality occurred in 1974. Based on

projections, the toll will reach three million in 1990.] Our wars go back to 1776. The traffic figure starts with 1900." The article concludes that "war is dangerous business but getting from one place to another by automobile is even more dangerous." Again, we should consider rates, not numbers, and comparisons should also consider the same age groups.

Peacetime versions of the same fallacy are also common. We often hear that "off-the-job activities are more dangerous than places of work, since more accidents occur off-the-job" or that "The bedroom is the most hazardous room in the home since more injuries occur in the bedroom than any other room." Here again the originators fail to consider differences in quantity of exposure, type of exposure, age, and other influencing factors.[3]

The confusion of correlation with cause.   In Chapter 7 "causes" of accidents will be discussed in detail. A favorite method of searching for "causes" is to hunt for statistical associations. It is often claimed, without real justification, that there are associations and correlations between contributing factors and accidents. Individuals often have difficulty contrasting the association of these factors with the total situation; illustrations are given in Table 3-2 to help them overcome this difficulty. The central point is that a genuine association exists *only* when two things appear together *either more frequently or less frequently than would normally be expected.*

Whenever two factors are associated, there are at least four possibilities as to why:

1. A causes B (epileptic seizure causes auto accident).
2. B causes A (auto accident causes seizure due to head injury).
3. Both A and B are caused by C (both seizure and auto accident are caused by flickering roadside lights).
4. A and B are independent and the "association" is by chance.

TABLE 3-2   When Does a Percentage Indicate a Statistical Association?

IF:	WE NEED TO KNOW:	BEFORE TRYING TO DECIDE:
50% of fatal accidents involve drinking drivers.	What percent of all driving is done by drinking drivers?	Whether drinking drivers contribute more or less than their share of fatal accidents.
50% of injuries to boys in the first three school grades are due to lack of safety knowledge.	What percent of all boys of similar age lack safety knowledge?	Whether lack of safety knowledge is associated with school injuries to boys.
30% of fatal accidents involve vehicles being driven too fast.	What percent of all driving is done beyond the speed limit?	Whether driving too fast is correlated with fatal accidents.
300% as many deaths occur off the job as on the job.	What percent of time is spent both on the job and off the job?	Whether a worker is safer at work than elsewhere.

The above shows how a statistical association never identifies the cause; it merely states that two factors move together, without indicating why. To explain what causes what, inferential statistics must be used, and this requires specialized training.

Sometimes an association or correlation is highly significant, even though the question of causes remains unanswered. For example, several insurance companies found that boys making good grades had fewer motor vehicle accidents than students making low grades. The company did not need to know the cause of this association, this fact alone was enough to permit them to cut premiums for boys who could show an average of "B" or better.

NOTES

1. *Accident Facts* may be obtained from the National Safety Council, 444 North Michigan Avenue, Chicago, Illinois 60611. Large public libraries may have copies for use.
2. William E. Tarrants, "Removing the Blind Spot in Safety Education Teacher Preparation," *School and College Safety*, National Safety Congress Transactions (Chicago: National Safety Council, 1965), p. 107.
3. *Ibid.*, pp. 107–8.

4

The Safety Movement

THE SAFETY TREND

Accidents have always plagued mankind. One of the earliest accounts of a concern for safety occurs in the eighth verse of the twenty-second chapter of Deuteronomy: "When thou buildest a new house, then thou shalt make a battlement for thy roof, that thou bring not blood upon thine house, if any man fall from thence." From this early admonition until the Industrial Revolution of the 1880s, accidents were the concern of the individual. The Industrial Revolution brought many changes—new hazards and new responsibilities which affected more people.

Factory inspections were introduced in England as early as 1833 and were designed to alleviate some of the worst hazards. But not until the twentieth century was any really effective attack made upon industrial hazards. Governmental regulations and controls were gradually formulated by most states in this country.

Effective labor legislation was passed between 1910 and 1915 and consisted of workmen's compensation laws. These laws required that the employer contribute to the costs of any work injury, whether or not a worker had been negligent.

Increased interest in safety also resulted in the formation of the National Safety Council in 1912. Initially formed out of concern for industrial safety, this agency was later expanded to include all aspects of safety and accident prevention.

We can see how effective the safety movement has been by examining accident rates for the past several decades. In general, statistics available for accidents indicate a definite decrease. However, the accident problem remains a significant one. There is still much room for improvement.

The National Safety Council reports that from its formation in 1912 to the present, accidental deaths per hundred thousand population have decreased over 50 percent, and if the rate had not decreased, nearly 2.5 million more people would have died as a result of accidents.[1] Of course, the success of death prevention is a result of the efforts of many organizations and individuals to alleviate accidental death.

We should be proud of the progress since 1912, but comparisons should be made for the past decade. Original concern focused on physical conditions. Since the 1930s, unsafe acts have been emphasized. Even with the progress shown in the total accident picture, we still need to reexamine our techniques for further success.

ORGANIZATIONS AND AGENCIES

Consistent and organized safety efforts have reduced the toll of accidents and their undesirable effects. Even so, they are still a major public health problem. Various organizations and agencies have emphasized to varying degrees accident prevention and injury control.

Since there are several hundred national organizations with the word safety in their title, the list below is by no means comprehensive in identifying those organizations and agencies with a specific interest in safety. In most cases the title of the organization identifies its main focus and interests. Anyone interested in a professional career in safety would do well to contact these and other safety organizations and agencies, remembering that the private sector of industry and business also employs thousands of safety specialists.

Government Agencies

U.S. Consumer Product Safety
  Commission (CPSC)
Washington, D.C. 20207

National Injury Information
  Clearinghouse
5401 Westbard Avenue — Room 625
Washington, D.C. 20207

National Highway Traffic Safety
  Administration (NHTSA)
400 7th St., N.W.
Washington, D.C. 20590

Federal Aviation Administration (FAA)
800 Independence Ave., S.W.
Washington, D.C. 20591

U.S. Coast Guard
Office of Boating Safety
Washington, D.C. 20590

Occupational Safety and Health
  Administration (OSHA)
200 Constitution Ave., N.W.
Washington, D.C. 20210

Clearinghouse for Occupational Safety
  and Health Information
4676 Columbia Parkway
Cincinnati, Ohio 45226

Division of Poison Control
Food and Drug Administration
5600 Fishers Lane; Room 18B-31
Rockville, Maryland 20857

Federal Emergency Management
Agency (FEMA)
Washington, D.C. 20472

U.S. Fire Administration
16825 South Seton Ave.
Emmitsburg, Maryland 21727

Mine Safety and Health Administration
(MSHA)
4015 Wilson Boulevard
Arlington, Virginia 22203

Professional Organizations

American Association for Health,
Physical Education, Recreation, and
Dance (AAHPERD)
1900 Association Drive
Reston, Virginia 22091

American Society of Safety Engineers
(ASSE)
850 Busse Highway
Park Ridge, Illinois 60068

American Academy of Safety Education
(AASE)
c/o Jack N. Green, Sr.
Safety and Driver Education
North Carolina A & T State University
Greensboro, North Carolina 27411

American Driver and Traffic Safety
Education Association (ADTSEA)
123 North Pitt St.
Alexandria, Virginia 22314

Other Organizations

National Safety Council (NSC)
444 North Michigan Ave.
Chicago, Illinois 60611

Insurance Institute for Highway
Safety (IIHS)
Watergate 600; Suite 300
Washington, D.C. 20037

National Fire Protection Association
(NFPA)
Batterymarch Park
Quincy, Massachusetts 02269

American National Red Cross
17th and D Streets, N.W.
Washington, D.C. 20006
or contact local chapter

American Automobile Association
(AAA)
8111 Gatehouse Road
Falls Church, Virginia 22042

National Rifle Association (NRA)
1600 Rhode Island Ave., N.W.
Washington, D.C. 20036

Motorcycle Safety Foundation (MSF)
P.O. Box 120
Chadds Ford, Pennsylvania 19317

American Trauma Society
P.O. Box 13526
Baltimore, Maryland 21201

To keep well informed and up to date on safety information and developments, referring or subscribing to the safety periodicals in Table 4-1 is recommended.

DISASTERS

Chapter 2 dealt briefly with the number of deaths resulting from disasters. As you will recall, the number of deaths from disasters is proportionately small when compared with other causes of accidental death; yet events called disasters receive front-page coverage in newspapers and news magazines.

The loss of life from disasters has decidedly influenced attempts to prevent needless death and injury. Table 4-2 presents a list of disasters with their respective death tolls and, more important, includes efforts made for reducing deaths and injuries following these disasters.

TABLE 4-1   Safety Periodicals

TITLE OF PERIODICAL	ISSUED	PUBLISHER	EMPHASIS
Professional Safety	monthly	American Society of Safety Engineers	occupational safety
Journal of Traffic Safety Education	quarterly	California Safety Association	driver education
Family Safety	quarterly	National Safety Council	all accident types
National Safety News	monthly	National Safety Council	occupational safety
Traffic Safety	bi-monthly	National Safety Council	traffic/motor vehicle safety
Journal of Safety Research	quarterly	National Safety Council and Pergamon Press	all accident types but primarily traffic safety
Accident Analysis and Prevention	bi-monthly	Pergamon Press	traffic/motor vehicle safety

FIGURE 4-1   Mount St. Helens Volcano. Photo courtesy of FEMA.

TABLE 4-2 Disasters and Their Effects on Safety Measures

TYPE	LOCATION AND DATE	TOTAL DEATHS*	RESULTS
Fire	City of Chicago, Illinois October 9, 1871	250	Building codes prohibiting wooden structures; water reserve
Flood	Johnstown, Pennsylvania May 31, 1889	2209	Inspections
Tidal wave	Galveston, Texas September 8, 1900	6000	Sea wall built
Fire	Iroquois Theatre, Chicago, Ill. December 30, 1903	575	Stricter theater safety standards
Marine	"General Slocum" burned East River, New York June 15, 1904	1021	Stricter ship inspections; revision of statutes (life preservers, experienced crew, fire extinguishers)
Earthquake and fire	San Francisco, California April 18, 1906	452	Widened streets; limited heights of buildings; steel frame and fire-resistant buildings
Mine	Monongah, West Virginia December 6, 1907	361	Creation of Federal Bureau of Mines; stiffened mine inspections
Fire	North Collinwood School Cleveland, Ohio March 8, 1908	176	Need realized for fire drills and planning of school structures
Fire	Triangle Shirt Waist Co. New York March 25, 1911	145	Strengthening of laws concerning alarm signals, sprinklers, fire escapes, fire drills
Marine	*Titanic* struck iceberg Atlantic Ocean April 15, 1912	1517	Regulation regarding number of lifeboats; all passenger ships equipped for around-the-clock radio watch; International Ice Patrol
Explosion	New London School, Texas March 18, 1937	294	Odorants put in natural gas
Fire	"Cocoanut Grove," Boston, Mass. November 28, 1942	492	Ordinances regulating aisle space, electrical wiring, flameproofing of decorations, overcrowding, signs indicating the maximum number of occupants; administration of blood plasma to prevent shock and the use of penicillin
Plane	Two-plane air collision over Grand Canyon, Arizona June 30, 1956	128	Controlled airspace expanded; use of infrared as a warning indicator

*Source: Based upon information from *Accident Facts* (1984), by permission of the National Safety Council.

The Galveston tidal wave of September 8, 1900, in which approximately six thousand lives were lost, was the most devastating disaster on record in the United States. A cyclone and tidal wave on November 12, 1970, struck East Pakistan, resulting in a disaster of proportions unprecedented in this century. The cost in lives from this natural disaster will never be reckoned accurately—a loss of more than half a million lives has been estimated.

The world's worst disaster on record (other than the Biblical account of Noah and the flood) occurred in 1887; a flood took 900,000 lives along the Hwang Ho River in China's Hunan Province.

TECHNOLOGY

We become aware of the impact of technology on safety and accident loss reduction by looking at various innovations. Whether the innovation be safety lenses (eyeglasses), safety belts (lap and harness types in automobiles), safety hats (steel or other hard material), safety shoes (steel-toed), or safety guards (on electrical saws)—all of which are widely accepted today—we can be sure that formerly they were viewed as quite unacceptable and met with widespread resistance.

LEGISLATION

Laws and regulations have resulted in a substantial reduction of injuries and fatalities. Examples can be given in air travel, railroads, boating, and mining, among others.

The case of motorcycle helmet laws in the United States portrays the effects of a law aimed at the reduction of death and injury. From 1960 to 1969 in order to qualify for safety program and highway funds, forty states enacted laws requiring motorcyclists to wear helmets. The law resulted in helmet use by more than 99 percent of motorcyclists. The law also had a dramatic effect on motorcyclists' fatalities. A study of eight states that adopted the law, compared to eight contiguous states without the law during the same period, found that motorcyclist deaths declined an average 30 percent following enactment of the law but did not change in the other states.

By 1975 all but three states had helmet-use laws that met the federal standard, but these three states did not lose any federal funds. Within three years, twenty-seven states changed their laws to allow most motorcyclists to ride without helmets. Comparison of helmet use and motorcyclist deaths before and after the repeal of the laws found use declined by more than half, on average, and deaths increased to rates that prevailed prior to the original enactment of the helmet-use laws.

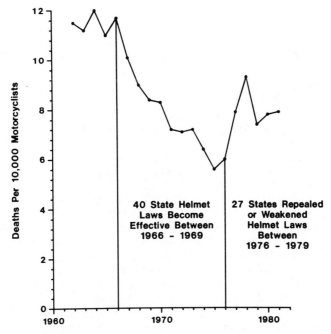

FIGURE 4-2 Motorcyclist deaths per 10,000 motorcycles by year, 1962-1981. From *The Injury Fact Book* by Susan P. Baker, Brian O'Neill, and Ronald S. Karpf (Lexington, Mass.: Lexington Books, D.C. Heath and Company, Copyright 1984, D.C. Heath and Company).

Major Safety Legislation

Despite tremendous gains in averting accidents and reducing death, injury, and property damage, the associated costs and losses to society still exist. Therefore, as an effort to combat specific accident types and injury losses, the federal government has passed several major safety laws. These laws include:

1. *The Highway Safety Act* was enacted to reduce motor vehicle injuries and fatalities by establishing a national highway safety program. This program required the establishment of uniform highway safety program standards upon which states and communities could organize their programs. Examples of standard areas include: periodic motor vehicle inspection, motorcycle safety, driver education, driver licensing, emergency medical services, pedestrian safety, and pupil transportation safety.
2. *The Occupational Safety and Health Act* was passed to assure, as far as possible, every working man and woman in the nation safe and healthful conditions and to preserve human resources. Thus, employers are required to provide a workplace free from hazards that are likely to cause death or injury to employees.

3. *The Consumer Product Safety Act* was created with the aim of protecting consumers against unreasonable risk of injury from hazardous products.
4. *The Federal Boat Safety Act* applies to all vessels used primarily for non-commercial purposes with the intention of creating a comprehensive boating safety program.
5. *The Federal Fire Prevention and Control Act* was enacted to reduce human and property loss through better fire prevention and control.
6. *The Federal Coal Mine Health and Safety Act* and the *Federal Metal and Nonmetallic Mine Safety Act* both regulate safety and health standards in the mining industry.
7. *Federal Hazardous Substances Act* was designed to require the labeling of hazardous products. A warning of the hazardous nature of a product in addition to any other pertinent safety information is to be presented on the container. Removal of products from sale that are hazardous and cannot be handled adequately through labeling is another feature of the law.
8. *Flammable Fabrics Act* was designed to ban the manufacture and sale of highly flammable fabrics used in clothing and interior home furnishings.
9. *Poison Prevention Packaging Act* was aimed at protecting people, especially small children, from various poisonous substances found around the house. A number of toxic and dangerous products are required to be sold in child-resistant packaging (e.g., aspirin, prescription drugs).
10. *Refrigerator Safety Act* was designed to protect against an individual's being trapped and suffocating inside a refrigerator by having it equipped with a device enabling the door to be opened easily from the inside.

Laws Directed at Individuals

There is a widespread use of laws and regulations which attempt to reduce accidents and their undesirable effects, and the imposition of new laws or regulations or changes in old laws or regulations often result in strenuous debate. The debate often focuses upon the right of the government to impose and enforce a law over the rights of individual freedom.

Although there are gaps in knowledge about the effects of laws and regulations directed at individual behavior, the evidence suggests some emerging principles regarding what works and what does not. H. Laurence Ross,[2] writing about deterring the alcohol-impaired driver, points out that laws have three basic components that separately and jointly can influence their effectiveness:

1. Likelihood of apprehension must be perceived to be high.
2. Swiftness of administering penalty must be assured.
3. Severity of penalty need not be extreme.

Most efforts aimed at influencing human behavior aim at making the law "tougher" by focusing on the severity of the punishment with little attention paid to the other two components. Ross points out that severe penalties by themselves do not deter offenders.

ECONOMIC COSTS

Only in recent years has the economic cost of accident prevention and injury control efforts been considered. Before, the cost of countermeasures seldom was an issue. In fact, some felt that a man's life was expendable. For example, when the Railway Safety Act was being considered in 1893, a railroad executive said that it would cost less to bury a man killed in an accident than to put air brakes on a railroad car.[3]

The struggle to eliminate or reduce the number of accidents and their undesirable consequences is based chiefly on two aspects: (1) cost and (2) regard for human life and well-being. It has been found that safety programs are good for business. Moreover, the first concern of almost everyone is for his or her health and safety. If these are safeguarded in one's everyday activities, the person is more productive and effective.

It has been estimated that the losses from accidental death, injury, and property damage are probably between $100 billion and $200 billion per year. A reduction in accident losses could therefore save billions of dollars annually.

Unfortunately, accident prevention and injury control programs are often undertaken because of compulsion or force from the government rather than for economic or moral reasons.

In recent years cost-benefit analysis has been a major consideration before implementing accident prevention and injury control countermeasures. The premise of cost-benefit analysis is that the benefits should equal or exceed the costs. Thus, there is concern when a countermeasure costs more than the social benefits.

The choice between two accident countermeasures that cost about the same but have different effectiveness is easy to make, as is the choice between countermeasures that have widely different costs but similar effectiveness. Unfortunately, safety decisions are usually not that simple.

MASS MEDIA

In their effects upon the daily lives of people, the mass media has probably changed more in the last fifty years than in all the preceding history of mankind. At the turn of the century, radio and television were unknown, the motion picture was a curiosity. Books were rarely available for other than the upper-middle and upper classes.

Since then, the mass media has become one of the most powerful influences on people's behavior and opinions. For example, though it is very difficult to document, television shows depicting car chases and crashes in which people seldom get injured could have a definite effect on people's driving habits. Television heroes rarely buckle safety belts, and magazine ads glamorize alcohol consumption. Such daily influences do not tend to promote a safer lifestyle.

Positive influences have resulted from the mass media as well, such as the showing of disaster films depicting the need for high-rise building evacuation plans or the acquisition of swimming skills. Books can also be powerful influences for change. For example, Ralph Nader's book, *Unsafe At Any Speed*, drew such national attention to a certain automobile model that the manufacturer made major modifications in the car's suspension system only to have the public already so acutely aware of the car's defects that sales dropped dramatically, resulting in the elimination of the car from production.

VALUES

How to spend one's life is answered only by one's choices among values. To this most important of all questions, science has no complete answer. Values are matters of preference, not matters of factual knowledge. Any attempt to say which values people ought to hold makes one a philosopher, not a scientist.

Nevertheless, almost all societies and cultures stress the value of human life and the preservation of self, family, and friends. Because this concept is almost universal, it is a major influence for emphasizing safety.

NOTES

1. National Safety Council, *Accident Facts* (Chicago: National Safety Council, 1984), p. 10.
2. H. Laurence Ross, *Deterring the Drinking Driver* (Lexington, Mass.: D.C. Heath and Company, 1982), pp. 102–4.
3. Willie Hammer, *Occupational Safety Management and Engineering* (Englewood Cliffs, N.J.: Prentice-Hall, Inc., 1981), p. 1.

5

Hazards, Risks, and Risk-Taking

The world seems to be a very hazardous place. Every day newspapers, radio, and television announce that some catastrophic accident has taken place. Everyone seems to be constantly subjected to an array of hazards.

The term "hazard" refers to a condition with the potential of causing death, injury, or property damage. Most hazards become dangerous only when humans interact with them. The term "danger" expresses a relative exposure to a hazard. A hazard may be present, but there may be little danger because of the precautions taken. Hazards must be identified before they can be controlled.

Hazards have always existed. Not all hazards are man-made. Technology does not always deserve the blame. Man has always had to contend with natural catastrophes, poisoning from plants, fire, and hypothermia. Although many of the old scourges have been conquered and in many ways we are better off than ever before, we still are confronted by new hazards such as radiation, electrocution, and automobile and airplane crashes.

The term "risk" refers to possible loss or the chance of a loss. Risk-takers often willingly expose themselves to hazards to obtain some possible gain, especially when in their personal evaluation, the possible gains outweigh the possible losses. Among the possible "rewards" or gains for successful risk-taking are: saving time, gaining status, experiencing a thrill, eliminating a hazard, meeting a dare or a challenge, and receiving a monetary reward.

Risk is a natural occurrence in human life. From an optimistic point of view, life is an adventure and man is continually pushing into new endeavors which are frequently dangerous. If there were no adventures nor excitement, no suffering or opposition, there would be no choice in life. Life itself would be rather monotonous. Without risk in our lives, no distinctively human kind of life would exist. Risk-taking in the sense of deciding to do one thing or another, is a critical and important part of being a human being.

From the moment we are born we are subject to hazards of all kinds. We continually run the risk of becoming injured or becoming a fatality in an accident. For each act of behavior, such as crossing a street or driving a car, there is an element of risk. The degree of risk may vary with time and situation; some risks are personal, some involve other people.

VOLUNTARY AND INVOLUNTARY RISKS

Risks can be subdivided in a variety of ways. One useful method separates voluntary and involuntary risks, although these terms have caused some confusion as to their meaning. In general, voluntary risk-taking involves some motivation for gain or benefit by the risk-taker. Involuntary risk-taking refers to the fact that the risk-taker does not have the opportunity to assess the benefits or options of his actions.

Someone who climbs mountains does so by choice and accepts whatever risk may be associated with the activity, whereas the risks arising from natural disasters, such as earthquakes or floods, cannot readily be avoided. The risk from mountain climbing is voluntary, while the natural disaster risks are involuntary. The distinction is not, however, always clear—guiding mountain climbers may be the only occupation open to a particular individual, so that the voluntary nature of his or her choice is qualified. Similarly, in principle at least, it is possible to move away from an area where the risk of floods or earthquakes is high. Nevertheless, the point is one to be kept in mind when considering the acceptability of risk.

INDIVIDUAL TENDENCIES TO TAKE RISKS

Any person who is directly faced with a dangerous situation makes his own judgments on risk-taking. When an individual feels he cannot, by himself, make a voluntary risk decision, the help of experts is often sought. Such experts might include physicians, safety professionals, or athletic trainers, who are especially competent to aid people because of their experiences and training. The degree to which the risk-taker considers the "expert" a "true" expert and the extent to which the advice is acceptable to the risk-taker determines how fully the advice is accepted. For example, if an ophthalmologist advised a worker to wear safety goggles to protect his eyes, the worker would tend to accept that opinion. But if a fellow worker suggested the same thing, the advice would have a lesser likelihood of being acceptable.

FIGURE 5-1 Mountain climbers accept whatever risks they encounter. Photo courtesy of Gerald Peterson, UDOT.

The degree of knowledge about hazards and risks falls into four classifications: (1) risk completely known to the risk-taker; (2) risk is hidden from the risk-taker; (3) risk information is easily available, but the risk-taker makes no attempt to use or acquire this information; and (4) risks are uncertain and undefined to all; no information is available.

An opinion about undesirable effects can change risk-taking behaviors. For example, an automobile driver may believe that an accident will seldom happen to him; then, immediately after witnessing a serious automobile accident, his opinion may change radically and he will seek to avoid an automobile accident by driving more cautiously than before, at least for a short time. Being too familiar with a hazard can lead to another view about the danger of a hazard. For example, most people try to avoid high tension power lines, but electrical linesmen tend to become less concerned by the danger as they daily work around the lines. Familiarity can generate a less cautious attitude.

FIGURE 5-2 Risk-takers make their own judgments about danger. Photo courtesy of Dan Poynter.

EXPRESSING RISK

There are three general methods of expressing risk: (1) those which apply relative methods; (2) those which use probabilities of occurrences of accidents; and (3) those which show a relation to exposure.

The relative method is simpler than the others, has many different forms, and is more widely used. The hazard is rated according to a standard. A rating scale may be adopted (1 to 10, 1 to 6, etc.) by which the risk level is assessed. For example, a liquid that has a flash point of 225°F may be more fire-safe than one that has a flash point of 150°F.

Probabilities of future occurrences of a specific type of accident can frequently be estimated from past experience if the experience has been over a long period of time, for a large population, and if the events to be assessed will occur under conditions similar to those under which the experience data were derived. The National Safety Council may predict that a certain number of persons will die in traffic accidents during a specific holiday. If the weather or the availability of gasoline should be different from that when the experience data were derived or from any assumptions the National Safety Council made, the actual number of fatalities will differ from the predicted number.

Table 5-1 gives a list of probabilities of an individual's having a fatal accident in the course of a year in certain types of accidents. The values shown were derived by dividing the total number of fatalities in the United States in a single year for a specific accident type into the total population that same year. This expresses the overall risk to society; the risk to any particular individual obviously depends on his exposure and a host of other factors.

Risks often are expressed in relation to exposure so that different risks can be compared. For example, in looking at different modes of transportation for passengers one might make comparisons in terms of fatality rates per passenger miles

TABLE 5-1 Probability of Having a Fatal Accident

ACCIDENT TYPE	NUMBER OF DEATHS*	CHANCE OF DEATH**
All accidents	105,000	1 in 2,157
Motor vehicle	52,600	1 in 4,300
Falls	12,300	1 in 18,400
Drowning	7,000	1 in 32,357
Fires, burns, and deaths associated with fires	5,500	1 in 41,181
Suffocation—ingested object	2,900	1 in 73,064
Poisoning by solids and liquids	2,800	1 in 80,892
Firearms	1,800	1 in 125,833
Poisoning by gases and vapors	1,500	1 in 151,000
All other types	18,400	1 in 125,833

*Source: National Safety Council, *Accident Facts,* 1981 (covers 1980 accidental deaths).

**Based on United States 1980 population of 226.5 million.

traveled. Thus, in a recent year the death rate per hundred million passenger miles was: for automobiles, 1.2 deaths; for commercial airplanes, 0.4 deaths; for passenger trains, .05 deaths; for public buses .04 deaths.

JUDGING SAFETY[1]

Safety is not measured; risks are measured. A thing is safe if its risks are judged to be acceptable. In other words, safety is the degree to which risks are judged acceptable.

The dictionary defines "safe" as "free from risk." Nothing can be absolutely free of risk. Everything thought to be safe, under some circumstances, can cause harm. Because nothing can be absolutely free of risk, nothing can be said to be absolutely safe. There are degrees of risk, and consequently there are degrees of safety.

Notice that this definition emphasizes the relativity and judgmental nature of the concept of safety. It also implies that two activities are required for determining how safe things are: (1) measuring risk and (2) judging the acceptability of that risk (judging safety), which is a matter of personal and social value judgment.

A false expectation is that safety experts can measure whether something is safe. They cannot because they can assess only the probabilities and consequences of events, not their value to people. Safety experts are prepared mainly to measure risks. Deciding whether people should be willing to take risks is a value judgment that the experts are little better qualified to make than anyone else.

Safety can change from time to time and be judged differently. Knowledge of risks evolves, and so do our personal and social standards of acceptability. For example, our decision whether to cross a street is different in different situations, depending on whether we are out in the rain, are carrying a heavy load of groceries,

FIGURE 5-3 Knowledge of risks evolves, and so do our personal and social standards of acceptability. Photo courtesy of Gerald Petersen, UDOT.

or are already late for an appointment. A power saw that is safe for an adult may not be safe in the hands of a child.

A risk estimate can assess the overall chance that a fatal accident will occur, but cannot predict any specific event. From past experience we can determine how many falls in the home will occur, but we cannot predict who, when, or exactly where the fall will occur.

Recall that determining safety involves two different kinds of activities: (1) risk-measuring—measuring the probability and severity of harm (a scientific activity); and (2) safety-juding—judging the acceptability of risks (a value activity).

The following table illustrates risk-measuring activities in the left-hand column, and corresponding safety-juding in the right column:

RISK-MEASURING	SAFETY-JUDGING
1. Bicycles rank first as a consumer product hazard (expressed by comparing).	1. Mandatory bicycle safety standards (i.e., reflectors, wheels, brakes, chains, etc.) have been issued by the U.S. Consumer Product Safety Commission.
2. Each American has one chance in 232 of being dog bitten each year (expressed by probability of occurrence).	2. Most communities have regulations about dogs being licensed and leashed, under verbal command, or confined.

3. Each American has one chance in 5,000 of being a traffic fatality (expressed by probability of occurrence).

3. The federal government has established standards in areas of traffic safety (i.e., alcohol control, driver education, enforcement, etc.)

How safe is safe enough is a value judgment. Scientists are not any more equipped to make value judgments about how safe things ought to be than any other person. It is dangerous to assume that those with technical or scientific expertise are especially qualified to make such judgments. Decisions, however, are often made by the experts because nobody else is prepared or willing to make them.

COPING WITH HAZARDS

When a risk-taker sees a new hazard, he has three options: fight, flee, or accept the imposed risk.

A risk-taker may seek to reduce his chances of injury by fighting the manner in which the hazard is imposed. For example, if an irrigation canal is scheduled to be constructed near a residential area and it poses the potential problem of having

FIGURE 5-4 Those engaged in risk-taking activities often spend many hours learning ways to minimize the hazards. Photo courtesy of the *Deseret News*.

young children drown in the canal, then residents of the area can attempt to have the canal redirected away from the homes or to have the canal fenced or covered.

A risk-taker may seek to reduce the chances of injury or death by removing himself from the hazard. For example, using the same problem cited above, a home owner could elect to sell his house and move to another location.

The third way of coping with a new hazard is to do nothing and accept the risk and hope that an injury will not occur.

PERSONAL GUIDELINES FOR RISK-TAKING

Risk is a relative thing. What may be risky to one person may be an everyday event to another. The margin of danger varies from person to person and even changes for the same person in different situations. In some risk-taking activities the margin of danger is the same for all. For example, in sky diving or mountain climbing, risk is obviously relative to the skill and experience of the participant—less skill increases the risk. However, in Russian roulette with a six-chamber revolver and only one bullet, the margin of danger is the same for all participants—there is no skill involved.

Most societies have their high risk-takers. Whether they be Vikings or astronauts, they have received a hero's reward. Sports such as bullfighting, football, snow skiing, and sky diving all involve risk. Today's heroes in these sports have spent many hours learning ways to minimize the hazards. Here are guidelines to consider when personally taking risks:

- Never risk more than you can afford to lose.
- Do not risk a lot for a little. Control emotions in a dangerous situation. Carefully identify what is the risk—a loss or a gain. Avoid risks merely because of a dare or for reasons of principle.
- Consider the odds and your intuition. There is a difference between taking a chance where there is no control and a situation where you are a major factor.

HAZARDS CONTROLLED BY TECHNOLOGY

As technology develops, new methods to control hazards are being found. These controls come in two forms: (1) hazards before considered uncontrollable can now be controlled to various degrees, and (2) more effective controls are found for hazards that are already controllable. An example of the first case is the research in progress on means of controlling tornadoes, either through cloud seeding or explosive disruption of the funnel. The degree of success of such controls is not yet established, but they are being tested. In the second case, new developments in building design (i.e., automatic sprinklers) are resulting in changes in building codes and, it is hoped, in better control of serious fires.

FIGURE 5-5
What may be risky for some may
not be risky for others. Photo
courtesy of Gerald Petersen,
UDOT.

FEAR TO RISK

Even though few people are killed in commercial aviation in the United States, there are twenty-five to thirty million Americans who are afraid to fly. These same people think nothing at all of driving their own cars without safety belts fastened, in spite of the fact that fifty thousand people will die in automobile crashes in a year's time.

There are three principles that govern perception of risk:

1. *Feeling in control.* People who drive "in control" or ski "in control" take risks, while a passenger on a plane feels little control over the situation.
2. *Size of the event.* Single big events (e.g., natural and human-caused disasters) are feared excessively and such events are often exaggerated by the media. One or two people dying in an automobile crash does not sound nearly as bad as forty-five dying in a plane crash.
3. *Familiarity.* It is hard to fear the familiar and hard not to fear the unfamiliar. For example, it is easy to fear elevators if you stay out of them; it is hard to fear them if you ride elevators several times every day.

TABLE 5–2 Risk-Taking

STEP #1:   Rate your participation in the following behaviors.
1 = frequently   2 = sometimes   3 = never

STEP #1 COLUMN		STEP #2 COLUMN	STEP #3 COLUMN
_____	1. flying in a small private plane	_____	_____
_____	2. flying in a commercial airliner	_____	_____
_____	3. swimming alone	_____	_____
_____	4. driving without safety belts on	_____	_____
_____	5. living on an active earthquake fault	_____	_____
_____	6. living in a "tornado belt" state	_____	_____
_____	7. working in an underground coal mine	_____	_____
_____	8. jay-walking across a street	_____	_____
_____	9. exceeding the 55 mph speed limit	_____	_____
_____	10. waving away or swatting at a bee	_____	_____
_____	11. tubing on a bumpy downhill course	_____	_____
_____	12. riding double on a bicycle	_____	_____
_____	13. petting or feeding a large stray dog	_____	_____
_____	14. keeping guns and ammunition together	_____	_____
_____	15. sleeping in a house without a smoke detector	_____	_____
_____	16. driving a small compact car	_____	_____
_____	17. taking pills or medicine in the dark	_____	_____
_____	18. talking with food in your mouth	_____	_____
_____	19. stopping incompletely at stop signs	_____	_____
_____	20. driving/riding a motorcycle	_____	_____

STEP #2:   For those statements that you marked as 1, try to identify one or more of the following reasons that explains why you participate.

    a. save time
    b. seek a thrill
    c. gain money
    d. meet a dare
    e. perform a necessary function
    f. gain recognition, status, attention
    g. eliminate a hazard
    h. other reason

STEP #3:   For those statements that you marked as 2 or 3 in Step #1, try to identify one or more of the following reasons that explains why you do not participate.

    1. It is not economical for me. It might cost me more than the personal benefits derived.
    2. It is too inconvenient. The time and hassle are not worth the benefits.
    3. It is too dangerous. An injury could happen.
    4. It does not provide enough psychological reward (i.e., thrill, recognition, etc.)
    5. I do not have enough skill to participate.

NOTES

1. The definition of safety and the concepts in this section are based on and are adapted from William W. Lowrance, *Of Acceptable Risk*.

6

Philosophical Implications

BASIC ASSUMPTIONS OF SAFETY

There are certain assumptions existing which aid in the understanding of our situation relating to safety in the world in which we live. These assumptions include:

- Hazards exist with the potential for causing death, injury, and property damage. Most hazards become dangerous only when humans interact with them.
- Basic laws and order of the physical world exist. We do not live in a chance world, but one governed by laws. For example, the solar system moves in mathematically correct orbits, enabling us to land a man on the moon. The earth's gravitational pull determines that what we throw into the air is pulled back to the earth; this always happens—even though the object might be a car that plunges off a mountain road taking the occupants to their deaths. Likewise, we know what will happen if we fall into a vat of boiling steel or come in contact with high tension electric power lines.
- Every law has both a punishment and a reward to it. There are opposites in all things. If not, the life known by humans would be mundane and monotonous.
- We are free agents, but we are prevented from doing what we want by limits set upon us by the environment in which we live, as well as by various social circumstances and limitations of the physical body. Being a free agent means that we can do what we please, but we will suffer the consequences of our actions—good or bad.
- Even though we are free to choose our course of action, we are not free to choose the consequences of our actions. Disobedience to law results in a

FIGURE 6-1 We live in a world governed by laws, not by chance. Exploration of the solar system and other accomplishments are possible when the laws are known. Photo courtesy of the National Aeronautics and Space Administration.

punishment. For example, if we touch a hot flame, we are burned. Obedience to law provides a reward. If we drive properly around a curve in the road, for example, we stay on the road.

- A knowledge of the basic laws can be acquired vicariously or through actual experience. It is important to realize that it is not necessary to break a law to learn what to avoid.

- To be adventurous is a natural aspect of human behavior and such adventure enriches life. In fact, progress depends on the adventurous urge (e.g., exploration which results in new discoveries).

- Because we are able to choose, we are responsible for our actions. "Responsibility" is here defined as the ability to fulfill our needs, and to do so in a way that does not deprive others of the ability to fulfill their needs. Acquiring responsibility is a complicated, lifelong process. Responsibility is learned through involvements with responsible fellow human beings.

- Because human lives are intermingled, what one person does may vitally affect someone else. When automobile accident insurance premiums rise, the owner of a car with an accident-free record becomes alarmed. After all, he or she is paying for others' accidents. Thus, other people's behavior affects us. When an automobile crosses through the freeway median guardrail and crashes into another car, killing its occupants, a stranger's actions have had a tragic effect on innocent people.

WORTH OF A HUMAN LIFE

What value is to be placed on a human life? How can we determine its worth? Two things will give some indication of the value of human life: (1) what these lives have cost up to this point—the labor, material, and struggle that has gone into their creation and development; and (2) the effective use to which they can be put—the benefits that result from productivity and contribution to society.

How much is a human life worth? All people have worth. Almost everyone recognizes a moral obligation to minimize the loss of life—most religions and philosophies are based upon the tenets that human life has worth and life is worth living.

The value of life often is measured in terms of money—how much is a company or a parent willing to pay to protect those under their stewardship from accidental death. For example, a parent may place a great deal of emphasis on safety by purchasing and using child restraints for their automobile, smoke detectors for the home, and appliances with UL ratings. Another parent may take none of these precautions.

There have been numerous ways in which the worth of a human life has been determined:

1. *Legal limitations.* Several states have set limits on the amount a victim's dependents can collect if someone else is held liable for his death in an accident. Other states have no limitations. Internationally, liabilities are sometimes limited by agreement among countries whose citizens could be involved.

Another aspect is that the life of a person who has no dependents usually has little or no legal value. Awards by courts are computed in a variety of ways. For example, a common method predicts the victim's total income had he lived to normal life expectancy, less his cost of living, and calculated on the basis of accrual of interest.

2. *Replacement method.* Replacing a productive, contributing life is costly. The military services have identified for the various military ranks the worth (in dollars) of individuals.

Another example is the replacement cost of housework of a woman who cares for her family full-time. Several methods have been used for such determinations, but suffice it to say, the calculations usually amount to a sizable amount of money.

Still another example is the loss to society by the premature impairment or death of persons with great talents and abilities (see Table 6-1). It is impossible to measure the intangible potential that was never realized because of an accidental death. Certainly the world would gain by increasing the length of life of talented people. In fact, the period of greatest achievement for artists, authors, scientists, and scholars occurs during mid-life according to several experts. Much depends, however, on such factors as the field of endeavor and the age at which creative work is begun. Most creative people maintain a high level of productivity throughout their lives. For example, the career of Thomas Edison reveals that, while the inventor's highest peak of discovery came at the age of thirty-five, he remained active and creative into his eighties.

3. *Insurance aspects.* The amount of insurance collected in the event of a death is another consideration.

4. *Accumulated assets.* Often appearing in the news media are listings of the richest people in the United States or in the world. Income tax collecting agencies (federal, state) appraise the worth of individual holdings at various times (e.g., probate court, bankruptcy, real estate values, etc.).

TABLE 6-1 Contributions Lost Due to Accidental Death

NAME	CONTRIBUTION(S)	TYPE OF ACCIDENT	AGE AT DEATH
Queen Astrid	Ruler of Belgium	Automobile crash	30
Albert Camus	French author and Nobel Prize winner for Literature	Automobile crash	47
Philippe Cousteau	Photographer, author, deep-sea diver and son of famous oceanographer	Seaplane crash	37
Pierre Curie	French Nobel Prize winner for Physics	Struck by truck while walking	47
James Dean	American movie star	Automobile crash	24
Princess Grace	Wife of Prince Rainier III of Monaco; former American movie actress, known as Grace Kelly	Automobile crash	52
Audie Murphy	American war hero; movie star	Small plane crash	46
Knute Rockne	Famous Notre Dame head football coach	Small plane crash	34
Will Rogers	American humorist, political analyst, movie star	Small plane crash	58
Percy Bysshe Shelley	English poet	Drowned while sailing during a storm	29
Natalie Wood	American movie actress	Drowned after falling from private boat	43

NOTE: Some gained fame that their names are instantly recognizable. Others are included because of noteworthy activities or achievements in their lifetime, even though their names are less well known.

BOX 6-1 *THREE FAMOUS PEOPLE WHO SURVIVED CRITICAL ACCIDENTS TO ACHIEVE SUCCESS*

1. Ben Hogan, golfer
   In 1949 he survived a near-fatal auto accident. After months of rehabilitation, he began playing golf again and the following year won the National Open.

2. Glenn Cunningham, track star
   As a youngster his legs were horribly burned in a fire. Doctors feared he would not walk again. However, he not only walked, but was the top miler in the world for many years.

3. Conrad Adenauer, German leader
   Almost killed in 1917 when he fell asleep while driving and crashed into a streetcar. He went through the windshield head first and his face was slashed beyond recognition. He survived, though his face was terribly scarred, and became chancellor of Germany after World War II.

5. *Value of the human body's chemicals.* Most people have heard that the human body is worth a few dollars on the basis of its minerals. However, the biologic chemicals found in human blood and tissue have an astronomical market value. Refer to Table 6-2 for examples of several compounds and their value.

6. *Social status.* Symbols of achievement and status are often used to place or determine someone's worth. Location of one's home, clothing labels, size and make of automobiles, type of employment, number of people at social functions (e.g., weddings, parties, funerals) all are examples of criteria used to determine one's worth.

Too often, we wrongly judge and evaluate a person's worth by material possessions rather than other standards such as contributions to society through service or creativeness.

TABLE 6-2   Value of the Human Body's Compounds

Using a chemical supply catalog, Daniel A. Sadoff, a University of Washington animal researcher, recently calculated the value of every marketable substance in a normal 150 pound body. Some of his findings appear below. Totaling the market values, Sadoff estimates we are each worth a million dollars.

COMPOUND	AMOUNT IN BODY (150 LB.)	VALUE
Cholesterol	140 g	$ 525.00
Fibrinogen	10.2 g	$ 739.50
Hemoglobin	510 g	$ 2,550.00
Albumin	153 g	$ 4,819.50
Prothrombin	10,200 U	$ 30,600.00
IgG	34 g	$ 30,600.00
Myoglobin	40 g	$100,000.00

DEATH

One of the most universal and solemn experiences shared by the human race is the inescapable fact of death. In fact, most of us hope for a long life and a quick death, but only a minority of the population has this wish fulfilled.

Two possibilities confront us: either we can die young and bring sorrow to family and friends, or we can live and feel sorrow as family and friends die. Such a statement is not pessimistic; it is simply true. For, sooner or later, in one way or another, all must die. How we as human beings act and react in the face of death has always been one of the principal concerns in life.

BOX 6-2

"I'm not afraid to die. I just don't want to be there when it happens."

— Woody Allen

FIGURE 6-2 A universal experience which comes to everyone is death. Photo courtesy of the *Deseret News*.

There seems to be an almost universal desire among men to continue to live. The desire for prolonging our lives is so strong, not only in human beings, but also in other forms of life, that the first law of nature is termed "the law of self-preservation." So deeply is the desire to live implanted in mankind that a human being will go to almost any length in order to preserve his own life.

BOX 6-3

> To everything there is a season . . . a time to be born, and a time to die . . .
> — *Ecclesiastes 3:1-2*

We seldom come in contact with the dead since few people die at home or in public places. Death takes place primarily in a hospital. However, some time in our lives we will be directly confronted by death.

It has been demonstrated that one's attitudes toward death can be improved (e.g., you can learn to accept death as a reality of living and be more comfortable with it).

Our attitudes toward our individual deaths affect not only the way we view death, but also the way in which we live our lives. For example, if one views his or her own death with horror, one may have considerable difficulty in mustering the courage necessary to cross a street in heavy traffic. If, on the other hand, death is conceived as a pleasurable and exciting experience, one may not hesitate to go over Niagara Falls in a barrel. Thus, an examination of death may provide clues as to our risk-taking behavior. The amount of risk-taking we are willing to do may be a function of our conception of death.

We can learn from death. To learn what is cold, we must learn what is hot. Likewise, is it not also logical that to learn about life, we must be aware of what death is?

SAFETY VALUES

Values are the things that people hold to be important. They are the things that people want or care about. Values are the guides that tend to give direction and purpose to live. Values represent a way of life.

In this book we are concerned with a particular value—the prevention of accidents and the control of death, injury, and property damage. We do make judgments about what is safe or unsafe. These judgments are value judgments.

The early slogan "Safety First," which made safety sound unattractive because it seemed to put safety ahead of every other consideration, came from an unfortunate circumstance. At the turn of the century, a man named Henry C. Frick, president of the Henry C. Frick Coke Company, inaugurated the country's first industry-wide safety campaign with the slogan, "Safety First, Quality Second, Cost Third." "Safety First" was lifted out of context and widely used by other companies. No longer was safety "first" only to quality and cost; the slogan seemed to put it ahead of everything.

We do *not* advocate safety as the main objective in life. We should not live by a timid and faint-hearted value. Safety is a positive concept, not a negative one. It is not meant to be a "thou-shalt-not" approach to life that holds people back.

If a person is to realize a useful and full life, the perception of safety as a positive value working with other individual and social values is essential. Safety as a value at times competes with time, status, personality shortcomings, bravery, adventure, and other values. Safety enables us to choose between experiences that are unproductive, absurd, and even stupid, and those that enrich our life, make it interesting and worthwhile. Safety is a means to an end—adventure, achievement of goals, progress (safety *for* as opposed to safety *from*).

Safety *for* enjoyment; safety *for* an effective life; safety *for* good health. These are some of the positive aspects of safety. The negative aspect, safety *from* death, injury, and property damage, should be discarded in favor of the positive approach. The safety *from* concept is only used with the very young, mentally deficient, and the senile elderly, since they are unable to understand the reason behind correct and safe procedures.

Safety *for* philosophy recommends activities with as much risk reduced as possible. Do not abandon snowmobiling, skiing, motorcycling, or any other activity simply because it has more risk than other activities—life then would become boring and unrewarding.

ISSUES IN SAFETY

Since safety is determined by the judgments or evaluation of people, a natural result is that people will not always agree about how safe they want to be. Refer to pages 6 and 43 for further elaboration on how safety is determined.

Since there are differing points of view with regard to various accident and injury problems, these differences are called issues. These issues should be studied because they:

1. Need resolving or answering.
2. Can and do affect people and property.
3. Are interesting and provocative.

Listed below are some controversial issues relating to accidental death and injury:

1. Mandatory use of safety devices such as automobile safety belts and motorcycle helmets.
2. Driver education's effectiveness in reducing accidents.
3. Handgun control.
4. Football and other contact sports in high schools.
5. Types of insulation used in housing.
6. Small vs. large cars for safeness in the event of an accident.
7. Efforts to control drunk drivers.
8. Boxing.

7

Determining Causes

MULTIPLE CAUSE CONCEPT

Accidents generally result from a combination of closely interwoven factors. This is known as the multiple cause concept.

Each of the circumstances which contributes to an accident is *a* cause, while *the* cause is the combination of these factors, each of which is necessary but none of which is by itself sufficient.

A *circumstance* is any condition or action accompanying an accident whether it contributes to the accident or not. A contributing *cause* is a circumstance without which the accident would not have happened. A cause is always a circumstance, but a circumstance is not always a cause. Each cause, if it truly contributes to an accident, is of equal importance in that accident. Too often the event directly preceding an accident is labeled the cause. The multiple cause concept refutes this.

We may "get away with" violations for years because all the other essential ingredients for the accident are not present. On the other hand, an accident could occur the first time the violation is committed.

One example of an attempt to show the multifactorial background of accidents is the chart shown in Figure 7-1 developed by the National Safety Council.

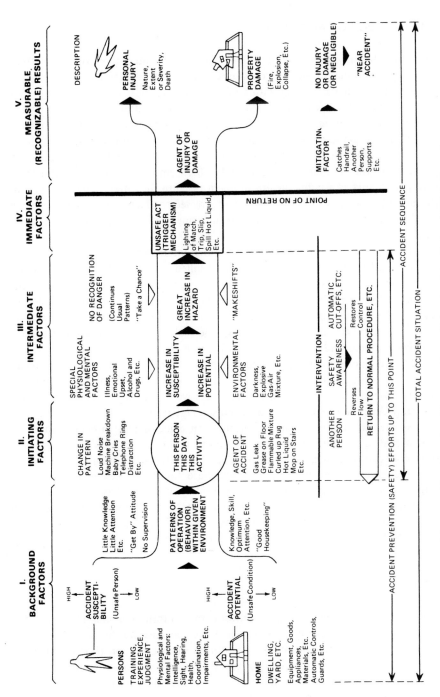

FIGURE 7-1 The dynamics of home accidents. Courtesy of the National Safety Council.

59

EPIDEMIOLOGY OF ACCIDENTS

"Epidemic" literally means "in or among people" and thus "common to, or affecting at the same time, many in a community." Formerly, epidemiology was defined as the medical science dealing with epidemics. It was most often found in public health programs that dealt with infectious and communicable diseases. However, heart disease, cancer, and accidents have also been studied by the epidemiological method. In relation to accidents, epidemiology is the study of why some get injured and some do not. John Gordon[1] (probably the first to use the epidemiological approach to the accident problem) and others emphasized the importance of viewing accidents as a public health problem because they play an increasingly significant role as a cause of death. They also stressed the fact that accident problems should be treated in the same way, with the same methods used in the epidemiological approach to other public health problems.

The epidemiological model includes the interaction of hosts, agents, and environmental factors. The *host* is the person killed or injured. The *agents* of injury are the various forms of energy in excessive amounts.[2] Mechanical energy is the usual agent for injuries involving motor vehicle crashes, falls, and gunshots. Thermal energy is the agent of burns. Chemical, electrical, and ionizing radiation are other forms of energy accounting for injuries. The lack of oxidation is the agent for drowning, choking, and other types of suffocation.

In epidemiological usage,[3] the term *vehicle* refers to any inanimate (nonliving) carrier that conveys damaging agents. The term *vector* is used to refer to animate (living) carriers. Most of the carriers of excessive energy are vehicles (e.g., cars, guns, etc.). A few are vectors (e.g., dogs, poison ivy, etc.).

No medical tool was more important than epidemiology in eliminating from Western civilization the old scourges of typhus, typhoid fever, cholera, plague, and many other diseases. Such can be the case for traumatic death and injury.

Precise knowledge is the only sound base for the implementation of accidental death and injury countermeasures. Safety programs based on sound epidemiological data must replace "hit-or-miss" safety programs. The fact that accidental death and injury are not often considered a public health problem is one of the reasons that they are indeed a public health problem.

DESCRIPTIVE EPIDEMIOLOGY

Descriptive epidemiology is by far the most common form of epidemiology utilized today. It provides statistical ways by which we can measure the accidental death and injury problem. By use of various sources of information (household interviews, report forms from hospital emergency departments, police reports, state and local health departments, insurance records, etc.) the injury problem is described in terms of frequency rates and characterized as to host, agent, and environmental factors.

The National Safety Council, the National Center for Health Statistics, the National Electronic Injury Surveillance System, and the National Highway Traffic Safety Administration operate systems for reporting the accident data obtained by their descriptive epidemiology efforts.

At first glance the goal of describing death and injury occurrence in this way may seem trivial. However, such studies are of importance and can serve the following purposes:

1. Focus attention on a particular accidental death and injury problem;
2. Measure long-range trends;
3. Serve as a basis for further studies and research;
4. Identify clues to accident causation;
5. Provide data to determine priorities for action;
6. Provide data to determine effectiveness of countermeasure programs.

Human

Human factors constitute a great concern in the descriptive epidemiology of accidents. Characteristics studied include age, sex, marital status, socioeconomic status, and physical condition.

Age. Age is one of the most important factors in accident occurrence. Some accidents occur almost exclusively in one particular age group, such as home fall fatalities. Other accidents occur over a much wider age span but tend to be more prevalent at certain ages than others.

The time of life at which an accident predominates is influenced by such factors as the degree of exposure to the agent at various ages and variations in susceptibility with age. The influence of age-related exposure is illustrated by lead poisoning, which is most prevalent in children.

Many injuries, such as in fatally injured adult pedestrians, show a progressive increase in prevalence with increasing age. It is tempting to regard an accident with this age pattern as being due merely to aging itself.

Current Age Tabulations. The tabulation of death rates in relation to age at one particular time, as in Table 7-1, is known as a current, or cross-sectional, presentation. This shows death rates as they are occurring simultaneously in different age groups; thus, different people are involved in each age group.

Sex. Some injuries occur more frequently in males, others more frequently in females. A sex difference in injury incidence initially brings to mind the possibility of hormonal or reproductive factors that either predispose or protect. For example, premenstrual syndrome (PMS) is a significant factor affecting women's susceptibility to accidents.

But men and women differ in many other ways, including habits, social relationships, environmental exposures, and other aspects of day-to-day living. The

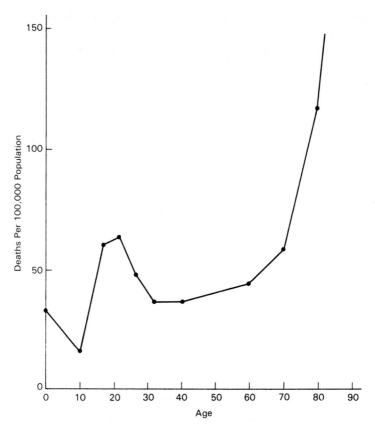

FIGURE 7-2 Injury Death Rates by Age. From *The Injury Fact Book* by Susan P. Baker, Brian O'Neill, and Ronald S. Karpf (Lexington, Mass.: Lexington Books, D.C. Heath and Company, Copyright 1984, D.C. Heath and Company).

TABLE 7-1 Accidental Death Rates by Age

AGE GROUPS	DEATH RATES*
All ages	38.9
Under 5 years	22.4
5 to 14 years	13.9
15 to 24 years	48.4
25 to 44 years	35.4
45 to 64 years	34.3
65 to 74 years	47.1
75 years and over	135.2

*Deaths per 100,000 population in each age group. Rates are averages for age groups, not individual ages.

Source: Deaths are for 1983, latest official figures from National Center for Health Statistics.

higher male prevalence of traffic fatalities is at least partly related to the fact that, on the average, men drive automobiles more than women.

Sex differences in injury occurrence are important descriptive findings and often suggest avenues for further research. Such differences need explaining. See Table 7-2 and Figure 7-3 for a comparison between males and females.

TABLE 7-2 Male: Female Ratios of Death Rates by Cause, 1977-1979

RATIO	UNINTENTIONAL INJURY	SUICIDE	HOMICIDE
52:1			Firearm (legal intervention)
29:1	Electricity (non-home)		
22:1	Fall from ladder/scaffold		
19:1	Machinery		
17:1	Struck by falling object		
12:1	Drowning (boat)		
10:1	Motorcyclist		
7:1	Fall from structure		
	Explosion		
	Lightning		
	Firearm		
6:1	Pedestrian (train)	Domestic gas	
		Firearm	
5:1	Drowning (non-boat)		Firearm
	Airplane crash		Beating
	Cutting/piercing		
	Electricity (home)		
	Bicyclist		
4:1	Opiate poisoning	Hanging	Cutting/stabbing
	Suffocation	Cutting/piercing	
	Motor vehicle exhaust		
3:1	Alcohol poisoning		
	Excessive cold		
	Motor vehicle occupant		
	Pedestrian		
	Fall from other level		
	Excessive heat	Motor vehicle exhaust	
2:1	Pedestrian (non-traffic)	Jumping	
	Exposure		
	Housefire		
	Aspiration (food)		
	Natural disaster		
	Aspiration (non-food)		
	Fall on stairs		
	Hot substance	Drowning	
1:1	Fall on level ———————————	———————————————————	————————
	Barbiturate poisoning	Barbiturate poisoning	
		Psychotherapeutic drugs	
1:2			Strangulation

Source: Susan P. Baker et al., *Injury Fact Book* (Lexington, MA: Lexington Books, 1984).

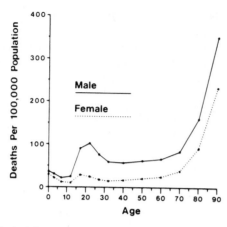

FIGURE 7-3 Death Rates from Unintentional Injury, by Age and Sex. From *The Injury Fact Book* by Susan P. Baker, Brian O'Neill, and Ronald S. Karpf (Lexington Books, D.C. Heath and Company, Copyright 1984, D.C. Heath and Company).

Marital Status is another important descriptive variable. Married persons have lower fatality rates than single persons. The unmarried have been shown to be more often involved in fatal traffic accidents than the married, and ski injuries are overwhelmingly found among the unmarried.

Socioeconomic Status. Some epidemiologists categorize socioeconomic status as an environmental factor. This factor can be and usually is measured by the occupation or income of the family head, by his or her educational level, or by residence, in terms of the value of the home or dwelling unit.

Low socioeconomic status appears to be related to lead poisoning, snowmobile injuries, and adult pedestrian deaths. On the other hand, drownings involving power boats and yachts, injuries or deaths related to private airplane crashes, and ski injuries are most often found among the high socioeconomic status groups.

Physical Condition. Although alcohol is among the most important human factors known to be related to severe injury and death, other drugs are becoming more frequently involved. The intoxicated have difficulty in escaping from a hazard (fire or submerged automobile), or intoxication becomes an obstruction that impedes those giving medical attention and treatment. Chronic health problems such as diabetes, epilepsy, and heart disease may also induce an accident. Overexertion and fatigue reduce the attention, sensory acuity, and reaction time of any individual.

Place

Where accidental injuries occur is a matter of great importance. Comparison of injury and death rates in different places may provide clues to causation or serve as a stimulus to further fruitful investigation. See Table 7-3 for a comparison between rural and urban death rates. Examples of descriptive findings presented are international comparisons, comparisons of regions within the United States, and comparisons of areas within a city.

TABLE 7-3 Rural: Urban Ratios of Injury Death Rates by Cause, 1977–1979

RATIO	UNINTENTIONAL INJURY	SUICIDE	HOMICIDE
10:1	Lightning; exposure		
9:1	Machinery; natural disaster		
7:1	Firearm		
	Struck by falling object		
6:1	Pedestrian (non-traffic)		
	Excessive cold		
	Drowning (boat)		
5:1	Suffocation		
	Motor vehicle occupant		
4:1	Electricity (non-home)		
	Explosion		
3:1	Fall on level		
	Clothing ignition		
	Airplane crash		
	Aspiration (non-food)	Firearm	
2:1	Alcohol poisoning		
	Electricity (home)		
	Motor vehicle exhaust		Beating
	Drowning (non-boat)		
	Cutting/piercing		
	Motorcyclist		
	Aspiration (food)		
	Fall from ladder/scaffold		
	Housefire		
	Bicyclist	Motor vehicle exhaust	
1:1			
	Hot substance		
	Excessive heat		
	Pedestrian		
	Pedestrian (train)	Drowning	
	Fall on stairs	Hanging	
1:2	Fall from structure	Psychotherapeutic drugs	Firearm (legal
		Cutting/piercing	intervention)
		Domestic gas	
1:3			Firearm
1:4	Barbiturate poisoning		
1:5		Barbiturate poisoning	
1:6			
1:7			Cutting/stabbing
1:10			Strangulation
1:20		Jumping	
1:29	Opiate poisoning		

Source: Susan P. Baker et al., *Injury Fact Book* (Lexington, MA: Lexington Books, 1984).

International Comparisons. Because of the problems regarding the validity of fatality statistics (attributable to definition problems), it is difficult to take seriously small differences among nations in accident fatality rates. However, it is also difficult to explain away very large differences (e.g., where the death rate is in one country two or three times as large as the death rate in another). Large differences are impressive when both countries are known to have reasonably good vital statistics systems.

Comparisons of Regions Within the United States. The availability of fatality statistics for states in the United States has permitted the discovery of interesting place-to-place variations in accidental death occurrence. Differences in fatality rates between urban and rural areas are a common finding. The higher fatality rate from traffic accidents in rural than in urban areas is consistent with the fact that faster driving occurs in rural areas, and it is the speed of the vehicle which is a major factor in death causation.

Geographic variation within the United States is quite distinctive, which suggests that climate or other factors may be involved (see Figure 7-4). For example, there is the finding in the United States of generally higher fatality rates for accidents in the mountain region. While hypotheses abound, to date no one has convincingly explained this geographic distribution of fatal accidents.

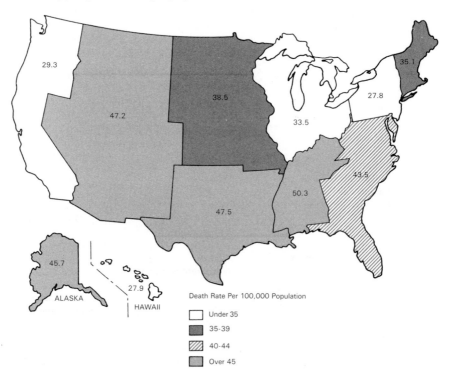

FIGURE 7-4 Accidental mortality by geographic division, United States. Adapted from the National Safety Council, *Accident Facts*, 1984.

Areas Within a City. When studying accidental injuries within a city, it is often desirable to plot the occurrence of such injuries in each census tract, since information about other characteristics of persons in each tract is available.

Lead poisoning, fire, and traffic accidents in cities have been mapped according to location with results showing definite geographic distribution.

Time

The pattern of injury occurrence in time is often an extremely informative descriptive characteristic. A great variety of time trends may be found in the accident data; these involve simple increases or decreases of injury incidence, or more complex combinations of these changes in time.

Short-Term Increases and Decreases in Injury Incidence. Short-term changes are those increases or decreases in injury incidence that are measured in hours, days, weeks, or months. For example, two-thirds of all drownings occur in the afternoon or early evening, and about 40 percent occur on Saturdays and Sundays. July is the peak month, with more than half occurring in the summer months, June through August.

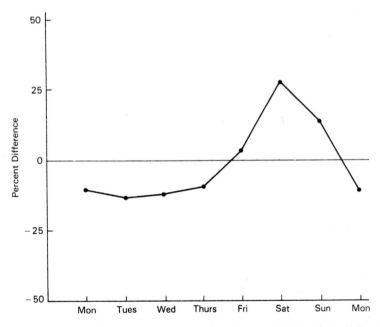

FIGURE 7-5 Percent Difference from the Average Number of Deaths from Unintentional Injury, by Day of Week. From *The Injury Fact Book* by Susan P. Baker, Brian O'Neill, and Ronald S. Karpf (Lexington, Mass.: Lexington Books, D.C. Heath and Company, Copyright 1984, D.C. Heath and Company).

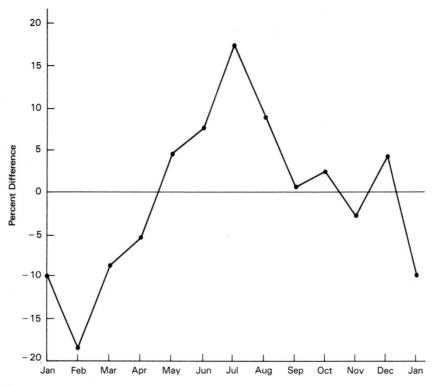

FIGURE 7-6 Percent Difference from the Average Number of Deaths from Unintentional Injury, by Month. From *The Injury Fact Book* by Susan P. Baker, Brian O'Neill, and Ronald S. Karpf (Lexington, Mass.: Lexington Books, D.C. Heath and Company, Copyright 1984, D.C. Heath and Company).

Epidemics. An epidemic, or outbreak, is the occurrence of an injury type in members of a defined population clearly in excess of the number of cases normally found in that population. Thus, the spectacular, short-term increases (some lasting several years) of injuries associated with backyard trampolines, skateboards, water ski kites, abandoned refrigerators, and ultrathin plastic bags were suddenly major, nationwide accidental death problems.

Recurrent or Periodic Time Trends. The incidence of certain accidental injuries shows regular recurring increases and decreases. This regular pattern exhibits cycles. Many cycles occur annually and represent variation in injury occurrence. Seasonal variation is a well-known characteristic for drownings, which occur mainly during the summer months, whereas carbon monoxide poisoning is most prevalent during the winter months when people are spending large amounts of time enclosed in automobiles and buildings.

Shorter-term periodic variations have also been observed. For example, death rates from automobile accidents show weekly cycles with the highest rates occurring on weekends, especially Saturdays. To date there are no available statistics on the number of passenger-miles driven on each day of the week. Thus, it is not possible to state whether the weekend increase in deaths is due merely to an increased exposure of the population to the moving automobile or whether the risk of death per passenger-mile actually increases, possibly because of such factors as more reckless driving or more alcohol consumption on weekends.

Long-Term Trends. Some accidental deaths exhibit a progressive increase or decrease in occurrence that is manifested over years or decades.

Figure 7-7 shows the death rates of accidents from 1910 to 1980. A marked decrease in fatalities from all accidents has occurred, representing about a 50 percent decrease. This decrease is believed to be due largely to a greater concern for safety. Rather than going back to 1910, it is suggested that only the past ten years should be considered.

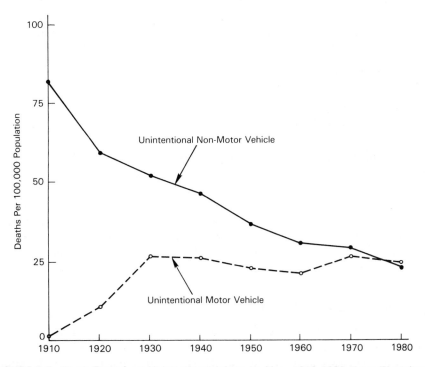

FIGURE 7-7 Death Rates from Unintentional Injury, by Year, 1910–1980. From *The Injury Fact Book* by Susan P. Baker, Brian O'Neill, and Ronald S. Karpf (Lexington, Mass.: Lexington Books, D.C. Heath and Company, Copyright 1984, D.C. Heath and Company).

INVESTIGATIVE EPIDEMIOLOGY

Investigative epidemiology is used to develop specific data related to the causes of the injury which would then point to feasible countermeasures. Whereas descriptive epidemiology identifies the accident problem and information regarding who, when, and where, investigative epidemiology gives the specifics on causation.

Complex interactions among the human, the environment, and the agent are disclosed. The investigations are conducted by a team of experts. These "multidisciplinary teams" involve a variety of professionals: medical/paramedical, social/behavioral, and physical/environmental personnel, engineers, statisticians, and clerks. Nurses, pathologists, toxicologists, psychologists, sociologists, and physical therapists have provided the types of skills needed. The effort is a teamwork approach, which overcomes the failings of an individual specialist.

Such investigations include: (1) studies of a single type of injury where a small number of cases can provide clues, (2) on-site investigations, and (3) in-depth evaluations.

Marland of the U.S. Public Health Service says this about the value of investigative epidemiology:

> The most important characteristic of investigative epidemiologic data is its specificity. When provided with the specific facts of injury-producing events, those persons responsible for injury control can develop an equally specific countermeasure. By the same token, those responsible for implementing the countermeasure are confident in their actions—be it redesign of a product, a local building code, a state law or an educational program. These programs became successful because their specificity is related to the injury just as surely as a cause and effect.[4]

It must be stressed that all accident investigations should seek causes for the injuries as well as causes for the accident, looking into all phases—before, during, and after—and all aspects of the scene—human, agent, and environment.

Present efforts by investigative epidemiology teams are small. Examples of such teams in operation include: Federal Aviation Administration (FAA) investigations of airplane crashes, Consumer Product Safety Commission teams interested in product safety, National Highway Traffic Safety Administration (NHTSA) team investigating motor vehicle accidents.

Obstacles to Acquiring Data

The multidisciplinary accident investigation teams have been encountering serious problems in obtaining information from people in the accidents they investigate. Witnesses or parties to the accident are often reluctant to divulge information. Despite the stated intention of the investigators to develop the data for research purposes only, parties confuse investigators with those seeking to bring legal actions against them or to involve them in such actions, both civil and criminal. As a result, people refuse access to information because they fear they will be harmed or embarrassed in some later legal action.

HUMAN CAUSATION MODEL

Many safety experts strongly believe that human error or behavior is the basic cause behind all accidents, and that safety is primarily a human problem. Furthermore, they believe that successes in reducing accident death and injury rates will come from controlling human behavior.

A number of writers have discussed and attempted to classify types of human errors, and a great deal has also been written about causes of human behavior. While some of these writers and theorists have psychology and management backgrounds, most come from the technical areas of systems engineering, systems safety, and human factors engineering.

Robert Mager and Peter Pipe[5] have identified human behavior problems. Using the flowchart in Figure 7-8, we start out by citing an unsafe behavior such as "Joe doesn't wear his safety belt," or "Tom is not storing his lawn mower's gasoline in a safe can."

The second step is to ask the question, "Is it dangerous?" The main idea here is whether the alleged unsafe behavior is really a problem.

Then, assuming the unsafe behavior is dangerous, we go to the next step, which is the determination of whether the unsafe behavior is due to a deficiency of safety knowledge and/or skills. In essence, is the person in danger because the person *does not know how* to be safe?

If the answer is "yes" we might determine whether the person ever knew how to perform safely. If there is a genuine lack of safety knowledge and/or skills, then the main remedy would be either to change the safety knowledge and/or skill level (e.g., teach the person how to be safe) or remove the person from the dangerous situation.

If, on the other hand, the answer is "no"—the person is able to behave safely but does not, the solution lies in something other than in enhancing safety knowledge and/or skills. "Teaching" someone to do what the person already knows how to do is going to change neither safety knowledge and/or skills nor safe behavior.

Here is an illustration: A plant manager complains that every year two million dollars are lost because of accidents. He believes the employees can recognize a hazard when they see one, but that they do not report hazards. The manager tried putting safety posters on the walls and had employees watch safety films regularly. Nothing happened to the accident rate because this is a case where people know how to be safe but are not. No amount of information or exhortation will change this situation. What is needed is a change in the conditions or the consequences surrounding employees' unsafe behavior.

Determining whether lack of safety knowledge and/or skills is due to a lack of training is one of the more important decisions in the anlysis of unsafe behavior,

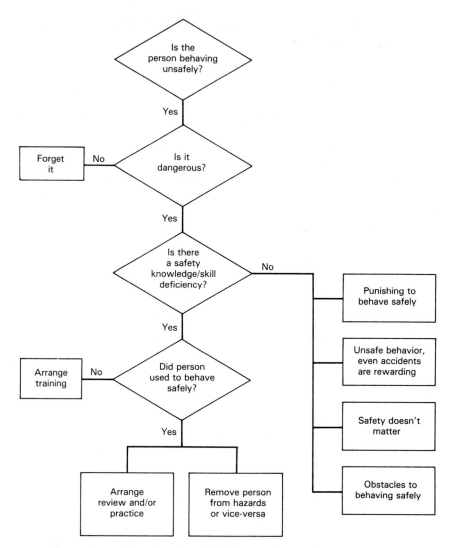

FIGURE 7-8 Human Causation Model. Adapted from Robert F. Mager and Peter Pipe, *Analyzing Performance Problems* (Belmont, Cal.: David S. Lake Publishers, 1984).

and is often neglected. If safety knowledge and/or skills never existed, training would be useful. But if safe behavior once existed and now is discarded, training from the beginning would be overdoing it.

When safe behavior fades or disappears and accident rates increase, periodic review programs should be considered. Such programs can include occasional review sessions of safety knowledge and/or skills or feedback about safe behavior from others.

To this point, several solutions have been suggested pertaining to the person who does not exhibit safe behavior due to deficiencies of safety knowledge and/or skills.

When a person could behave safely if he really had to but does not, something other than safety instruction is needed. There are four general causes of such unsafe behavior:

1. It is punishing to behave safely.
2. It is rewarding not to be safe, or it is even rewarding to have accidents.
3. Safety does not matter.
4. There are obstacles to being safe.

When it appears that someone knows how to be safe but does not act that way, find out whether safe behavior has been leading to unpleasant results. Some examples of how it is punishing to act safely include: (1) It is time consuming or a waste of time; (2) it is inconvenient; (3) it is uncomfortable (e.g., helmets and goggles are hot); (4) an attitude to not "rock the boat" by reporting hazards exists; (5) it is considered unmanly ("sissies" use goggles or a guard); (6) it is difficult to use equipment because of guards, etc. If so, the remedy is to find ways to reduce or eliminate the negative effects and to create, or increase, positive or desirable consequences.

Secondly, there are cases where the consequences of unsafe behavior are more favorable than those that follow safe behavior. Examples of how unsafe behaviors are rewarded include: (1) injured victims gain attention and (2) those having been injured get to rest, get time off, etc. If so, the remedy is to find ways to reward those who are practicing safety.

Thirdly, sometimes unsafe behavior continues to exist not because the person does not know how to perform or because the person is not motivated, but because it simply does not matter whether the person behaves safely. Nothing happens if he is safe; nothing happens if he is unsafe. In a case where a person could be safe if he had to or wanted to, one of the things to look for is the consequence or benefit of being safe. If there is none then it may be necessary to arrange one. When you want someone to behave safely, one rule is to "make it matter."

Lastly, if a person knows how to be safe but does not behave that way, look for obstacles. Look for things that might be getting in the way of his safe behavior. Look for lack of time, lack of authority. Look for poorly placed or labeled equipment. Look for bad lighting and uncomfortable surroundings. Most people want to avoid accidents. When they do not, it is often because of an obstacle in the environment around them.

Once the reason is found for unsafe behavior, it is worth determining a worthwhile remedy or solution. Select from several solutions the most practical, economical, and easiest to use—the one most likely to give the most result for the least effort. This is cost effectiveness.

SYSTEMS SAFETY APPROACH

The engineering techniques of systems analysis have also been proposed for the solution of safety problems. In its emphasis on the interaction of humans, machines, and environment, this approach is quite similar to the epidemiological one.

Too many well-written explanations of the systems safety approach have been published for this writer to elaborate further. Suffice it to define systems safety for the uninitiated with the hope that he or she will investigate further. Systems safety analysis is a logical, step-by-step method of examining situations with accident potential It helps identify causes that could be overlooked if a less detailed method were used. Human, mechanical, and environmental factors are placed in an orderly, related schematic, inducing a logical thought process which leaves little to chance. It is a challenging and satisfying creative brain-storming approach.

While the systems safety approach has been highly successful in regard to the safe operations of some very complex systems, many problems yet remain in quantifying the influence of the human as an element in the system.

A system is an orderly arrangement of components which are interrelated and which act and interact to perform some task or function in a particular environment. The main points to keep in mind are that a system is defined in terms of a task or function, and that the components of a system are interrelated, that is, each part affects the others. Examples of systems are aircraft, production lines, and transportation.

No matter which method of analysis is used, it is important to have a model of the system. Most models take the form of a diagram showing all the components. This makes it easier to grasp the interrelationships and simplifies tracing the effects of malfunctions.

The systems approach to safety can help to change safety from an art to a science by classifying much of our knowledge. It can change the application of safety from piecemeal problem solving to a safety-designed operation. We can apply the question "What can happen if this component fails?" to the various elements of the systems and come up with adequate safety answers before the accident occurs, instead of after the damage has been done.

OTHER HUMAN FACTORS

Accident Proneness

Some people do have more accidents and injuries than others. In fact, a mere look around at friends and neighbors will usually identify several who have had more than their share of injuries.

Accident proneness refers to the existence of an enduring or stable personality characteristic that predisposes an individual toward having accidents. Critical words in this definition are the adjectives *enduring* and *stable*.

The concept of accident proneness has had a long and controversial history. English researchers were responsible for the early work (1919) which *created* the concept, which then *grew* in popularity through the 1920s, reaching its *peak* in the

BOX 7-1

> He is so unlucky that he runs into accidents which started out to happen to somebody else.
>
> — *Don Marquis*

late 1930s. Accident proneness was *questioned* in the 1950s and *rejected* in the 1960s. Since then, it made somewhat of a *comeback* in the 1970s. Needless to say, the accident proneness concept has been plagued with argument, debate, and confusion. This has largely resulted because the concept of accident proneness means different things to different people.

Misconceptions which have been associated with accident proneness include:

Most accidents are caused by a few people. This point of view has been represented by "20 percent of the people who have 80 percent of the accidents." The 20 percent were identified as being accident prone.

The major flaw in this concept is the fact that people who have accidents in one period of time are not necessarily those who have accidents in the next. That is, 20 percent of a population may indeed have 80 percent of the accidents during a two-year period, but during the next two years the 20 percent of the population involved in accidents does not include all of the original individuals. Thus, uneven accident involvement may best be explained on the basis of chance alone, and there is no reason to predict that an individual in the "20 percent group" will continue to be heavily involved in accidents.

Accident proneness as a single personality trait. Confusion arises because some authorities consider accident proneness as a single trait, while others will include in the accident prone definition a variety of psychological traits.

There is little agreement on what traits distinguish the accident prone personality. In fact, the descriptions often conflict. For example, one study found the accident prone to be passive and repressed, while another described him as impulsive and action oriented.

Accident proneness as an influence found in all environments. This view is depicted by the well-known quote, "a man drives as he lives," indicating that an "accident prone" person would be susceptible not only for a traffic accident but an accident in other surroundings as well (i.e., work, home, etc.).

Accident proneness as an innate and unmodifiable trait. Such a belief stigmatizes an individual and would be contrary to the popular idea that behavior can be modified. There are no known psychological traits that predict the probability of an accident with any accuracy.

People do change. Maturing with age and becoming married are examples of experiences which have influenced and even changed an individual's behavior and character.

Accident repeaters are accident prone. Accident repetitiveness refers to the *descriptive* truth that some individuals have more accidents than others. Accident proneness is a theory and is offered as the *explanation* of why some people have more accidents than others. Evidence indicates that repeaters are not necessarily prone and that exposure to hazards may account for many of those having more than their share of accidents.

One of the reasons for the failure of the accident proneness theory has been the lack of agreement over its meaning. Accident proneness means different things to different people.

The bulk of evidence indicates that there is no such thing as a single type of "safe" or "unsafe" person. Rather, each individual has a range of behavior—some safe, others unsafe, depending on the environmental hazards to which he is exposed. Thus, instead of a single type of accident prone individual, there are many reasons why some individuals have more accidents than others.

For example, individuals vary considerably in their: (1) exposure to hazards, (2) ability to recognize and make judgments concerning hazards, (3) experience and training, (4) physical attributes (i.e., visual acuity, reflexes, etc.), and (5) exposure to social and environmental stresses.

The idea of accident proneness must be viewed as only one explanation for individual variations in accidents. There are other explanations which must always be considered.

Criticism of the accident proneness concept should not be taken to mean that personal factors do not play an important role in accidents. Categorization of human factors involved in accidents may be divided on a short-term and long-term basis.

BOX 7-2

My only solution for the problem of habitual accidents . . . is for everybody to stay in bed all day. Even then, there is always the chance that you will fall out.

— Robert Benchley

Type I—Short-Term

Emotional. This person is under stress. The stressful period can last anywhere from several weeks to several months and it may occur only a few times in a life-time or several times during a year. During this period, the individual is likely to show forgetfulness, inattention, unusual irritability, and other symptoms. Once the crisis is resolved, the individual goes back to his previous state of good adjustment.

Physical. A person ill or recovering from an illness, fatigue, using alcohol or other drugs may be involved in a period of susceptibility for an accident.

Type II—Long-Term

Emotional. In this category are individuals with "negative character traits" such as untrustworthiness, rudeness, and aggressiveness. This category reflects the usual definition of accident proneness—that of having a personality trait enduring over time, perhaps a lifetime. However, a person may change his character as the result of becoming more mature with age, education, the need to be approved, marriage, responsibility, or simply "slowing down" from old age.

Physical. People who are suffering physical conditions which may impair their ability to perform safely, such as failing eyesight, senility, untreated diabetes, seizures not under drug control, and a host of other physical conditions, may account for accident susceptibility. However, many people with known chronic ailments compensate for their condition and do perform safely, and, in fact, may even have better accident records than those who are healthier.

Lack of Knowledge and Skills

With all the new inventions coming forth on a daily basis, no single individual can keep up with all the new innovations. If the innovation happens to be hazardous, people will get hurt in many cases merely because they did not know how to handle the new contraption. Examples of hazardous products later deemed unsafe include hot water vaporizers capable of severely burning when tipped over, electrical toy ovens with the potential of high-level temperatures, and power lawn mowers capable of ejecting foreign objects.

Another problem exists when a person could use a potentially hazardous object safely yet because he or she does so infrequently, the person endangers himself or herself.

Imitation

People are strongly influenced by imitation. "Do as I say and not as I do" is a recognition of the fact that people sometimes learn more from imitation than we would like. Children acquire much of their behavior by watching parents and peers. Often they imitate behavior new to them following only brief exposure to it. Research confirms the fact that when you teach one thing and then do something else, the teaching is less effective than if you practice what you teach. Safe driving and use of safety belts are both examples of imitation's effects. We tell parents, "do not take medications in front of your children." The reason: fear that they will imitate the parent and become poisoned.

Attitudes

It would be difficult to overemphasize the importance of attitudes relating to accidents. As is evident to most people, attitudes are of major significance in determining how safe a person might be.

An attitude is an internal state which affects an individual's choice of action toward some object, person, or event. The choice an individual makes is of a personal nature. For example, a person may choose to drive faster than the speed limit or use a power saw without a guard—the choice indicates his or her attitude.

Accidents may even affect attitudes. Certain beneficial attitudes may come from accidents. An involvement in an accident or witnessing an accident may foster greater respect for laws, social responsibility, and moral obligation. Thoughtfulness, courtesy, sportsmanship, and other qualities may sometimes result from accidents (e.g., an individual who causes an injury to another may become more thoughtful of others in similar future situations). There may also be negative attitudes resulting from accidents. Emotional disturbance, nervousness, anxiety, and other psychological problems may result from accidents. Fear, anger, hatred, and antisocial behavior may be also produced as a result of being involved in an accident.

Attitudes are a fundamental factor in safety. Attitudes stem from observation, learning, and experiences of an individual. It is often difficult to modify or change attitudes since they usually have been developed over a long period of time; nevertheless, attitudes can be altered toward a safer lifestyle.

Alcohol and Other Drugs

Alcohol and drug abuse are behaviors with major implications in all areas of safety. More than half of the fatal motor vehicle accidents and up to 40 percent of adults dying in non-motor vehicle accidents (e.g., poisoning, falls, burns) had used alcohol. Reports indicate that almost half of all adults who drowned had consumed alcohol.

Drugs other than alcohol have not been shown to play a substantial role in accidents, although abuse of amphetamines, marijuana, or other drugs can seriously impair performance. Multiple drug use, especially the combination of alcohol with one or more other drug, creates additional problems.

Emotions

Little information in this area is available. However, the likelihood of accidents increases when people are emotionally upset. An emotional state (e.g., anger) can carry over the manner of driving or other behavior which involves hazards.

BOX 7–3   *UNUSUAL BUT TRUE*

NEW ORLEANS (AP)—Mourners gathered Friday for Michael Scaglione, 26, who was killed after he threw a golf club and the clubhead rebounded and stabbed him in the throat, severing the jugular vein.

Police said Scaglione made a bad shot and threw his club against a golf cart. When the shaft broke and the head rebounded, he was struck in the throat.

Other members of the foursome said Scaglione staggered back, gasped, "I stabbed myself," and pulled the piece of golf club from the wound.

Surgeons said if Scaglione had not done that he might have lived, since the metal might have reduced the gush of blood. Rushed to a hospital, he was revived temporarily, but died later.

Nearly all people have periods in which they experience emotional stress. During these periods they will tend to be less alert and less attentive to what they are doing, and there will be a tendency to have accidents.

Various types of emotions may contribute to accidents. Extreme fear, anger, nervousness, or anxiety are examples.

Fatigue

Evidence shows that as persons become tired they are more likely to experience accidents. The true magnitude of the fatigue problem is not known; however, it is a larger problem than accident data indicates in all areas of safety—motor vehicle, school, work, etc.

There are basically two types of fatigue: "task or skill induced" fatigue and fatigue from other factors, of which sleep deprivation is the most common. "Task or skill-induced" fatigue results from long hours of doing essentially the same thing resulting in loss of alertness and difficulty in responding properly to hazards.

Stress

The old concept of "accident proneness" (e.g., a lifelong personality trait) has been largely abandoned today as an explanation of accident causation and involvement. Instead we should be looking at the acute situational factors which may precipitate an accident. In fact, accident involvement often can be explained by personal stresses which cause a person to perform in such a manner as to increase his or her accident chances.

Everyone has experienced stress; moreover, we have some degree of stress all of the time. Everyone has a different level of stress tolerance, which means each individual is capable of handling different amounts and kinds of stress. However, everyone has his or her limits.

Society's Values

The American values taught in homes and schools include:

- Competitiveness in an aggressive style is acceptable.
- Masculinity implies toughness and aggressiveness.
- Excitement and challenges should be sought, and risk-taking is justified in seeking them.

These are the values exemplified by our historical heroes. These beliefs and others have allowed the United States to excel in all areas (e.g., sports, finance, technology, etc.). All of this is relevant to accidents because these values suggest that exposure to hazards in a certain type of lifestyle is the ideal. Such a lifestyle lends itself to involvement in accidents.

Biorhythm

The biorhythm theory proposes that many accidents occur on certain critical days, which are calculated from a person's birth date.

The biorhythm concept was developed in the early 1900s by Wilhelm Fliess, a Berlin physician, and Herman Swoboda, a Viennese psychologist. Working separately they concluded that each of us has a 23-day physical cycle and a 28-day emotional cycle. Late in the 1920s, Alfred Telscher, an Austrian engineer and teacher, added a 33-day intellectual cycle after observing his students' academic performance. The work of these three men forms the basis of today's biorhythm theory.

The use of biorhythm appears to be widely exaggerated—at least in the United States. Only a few organizations will admit using the theory, and mostly to promote safety awareness, not to predict accidents.

The biorhythm cycles are usually depicted as curves, with a positive phase, a negative phase, and a zero or "critical point," when the lines cross the axis from positive to negative, or vice versa. Although negative days are regarded as less favorable than positive days, it is the critical days—occurring about six times a month—that are considered most upsetting.

Opinions on the validity of biorhythms vary. Some claim the theory has been successfully used in accident prevention programs by companies in the United States, Japan, and Europe, while others say the research results are inconclusive.

There are other types of biological cycles. For example, the strongest cycle is the 24-hour circadian rhythm. Various body functions that move in a 24-hour pattern are impressive: temperature, blood pressure, respiration, blood sugar, urine volume, coordination, and many others.

When the circadian rhythm is upset, problems can result. Everyone is familiar with the fatigue and disorientation that results from a long period without sleep. Other examples of upsetting the rhythm include "jet lag" in travelers flying long distances across several time zones, and shift work on jobs.

Physical and Medical Conditions

Medical conditions and physical defects are frequently associated with accidents. The following items are not all-inclusive list, but they have been indicted as causes of accidents: circulatory problems, diabetes, narcolepsy, premenstrual syndrome, epilepsy, vision and hearing defects, among many others.

NOTES

1. John E. Gordon, "The Epidemiology of Accidents," *American Journal of Public Health*, XXXIX (April 1949), 504–15.
2. William Haddon, Jr. "Advances in the Epidemiology of Injuries as a Basis for Public Policy," *Public Health Reports* 95, no. 5 (September-October 1980), 412–13.
3. *Ibid.*, p. 413.
4. Richard E. Marland, "Injury Epidemiology," *Journal of Safety Research*, (September 1969), p. 102.
5. Robert F. Mager and Peter Pipe, *Analyzing Performance Problems* (Belmont, Calif.: Fearon-Pitman Publishers, Inc., 1970).

8

Strategies for Accident Prevention and Injury Control

CHOOSING COUNTERMEASURES

Measures to prevent accidents and reduce injuries have been in existence since ancient times (for example, shoes to protect against sharp stones). Many worked so well (such as evacuation in times of floods and volcanic eruptions) that they are still in use.

Traditional, often ineffective, approaches to safety have been based largely on piecemeal, unsystematic perceptions of accident causes and countermeasure options. Utilization of the criteria listed below will aid in determining countermeasures:

1. Priority should be given to countermeasures that will be the most effective in reducing injury losses and need not be based on causes which contributed to the accident. Some examples of this would be: using a net for an acrobat rather than urging him or her to "perform safely" and not to fall; placing thermal insulation on the handles of cooking pots, pans, and electric irons rather than telling users never to touch them with their bare hands; putting shoes on children rather than cautioning them not to stub their toes.
2. A mixture of countermeasures is best since a single measure will rarely solve the problem.
3. Quickness in obtaining positive results from a particular countermeasure is important.

4. Economic factors should play a major role in the choice of options. The often-heard statement that "it's worth it, if it saves just one life" is not true if, for the same amount of money or other resources, more than one life can be saved. Cost-effectiveness compares the cost of alternative ways of meeting a particular goal, whereas cost-benefit analyses assess the net financial gain or loss to society of implementing a particular safety program. Cost-effectiveness and cost-benefit analyses are often misapplied and misunderstood in decisions concerning safety programs.

Cost-effective designs should be chosen to minimize societal costs. Because of difficulties in determining the value of life and limb, cost-benefit studies would be appropriate only in the evelution of countermeasures intended to reduce property damage, but not save lives and reduce injuries.

5. In most cases, the less that the people to be protected must do, the more successful the countermeasure. Automatic ("passive") measures do not require the person to do something, and manual ("active") measures require that they do.

6. There is nothing sacred about any specific model for determining countermeasures. Whichever model is used should serve as a sorting device, a checklist that assists the user.

7. Consideration of the cultural, social, and political forces will sometimes determine success or failure.

8. Choice must be based on effectiveness in helping to reduce the end results (death and injury)—not always on preventing an accident.

The following poem shows the importance of stressing one countermeasure over another:

The Parable of the Dangerous Cliff

Twas a dangerous cliff, as they freely confessed,
  Though to walk near its crest was so pleasant;
But over its terrible edge there had slipped
  A duke, and full many a peasant.
The people said something would have to be done,
  But their projects did not at all tally.
Some said, "Put a fence 'round the edge of the cliff,"
  Some, "An ambulance down in the valley."

The lament of the crowd was profound and was loud,
  As their hearts overflowed with their pity;
But the cry for the ambulance carried the day
  As it spread through the neighboring city.
A collection was made, to accumulate aid,
  And the dwellers in highway and alley
Gave dollars or cents—not to furnish a fence—
  But an ambulance down in the valley.

"For the cliff is all right if you're careful," they said;
  "And if folks ever slip and are dropping,
It isn't the slipping that hurts them so much
  As the shock down below—when they're stopping."

So for years (we have heard), as these mishaps occurred
  Quick forth would the rescuers sally,
To pick up the victims who fell from the cliff
  With the ambulance down in the valley.

Said one, to his plea, "It's a marvel to me
  That you'd give so much greater attention
To repairing results than to curing the cause;
  You had much better aim at prevention.
For the mischief, of course, should be stopped at its source,
  Come, neighbors and friends, let us rally.
It is far better sense to rely on a fence
  Than an ambulance down in the valley."

"He is wrong in his head," the majority said;
  "He would end all our earnest endeavor.
He's a man who would shirk this responsible work,
  But we will support it forever.
Aren't we picking up all, just as fast as they fall,
  And giving them care liberally?
A superfluous fence is of no consequence,
  If the ambulance works in the valley."

The story looks queer as we've written it here,
  But things oft occur that are stranger.
More humane, we assert, than to succor the hurt,
  Is the plan of removing the danger.
The best possible course is to safeguard the source;
  Attend to things rationally.
Yes, build up the fence and let us disperse
  With the ambulance down in the valley.[1]

CHANGING HUMAN BEHAVIOR

It is generally agreed that a majority of accidents are caused or at least greatly influenced by human behavior. Therefore, attempts to prevent accidents focus upon changing human behavior by one or more of the following ways. It should be noted that such attempts usually begin with education and continue down through the others if the preceding one is not effective. For example, if education fails, then coercion is tried; when that fails, legal sanctions are endorsed. This whole process may take years or decades before an effective countermeasure is found or developed.

Education. Organized safety education is of recent origin and it is impossible to pinpoint the origin of the first safety instruction taking place. It is known that *McGuffey's Readers* included a number of references to safe practices. Since then many forms of safety education have appeared for all accident types.

An evaluation of any safety instruction should answer at least the following questions concerning a safety lesson, topic, or course: (1) To what extent have the stated objectives of instruction been met? and (2) Is the safety instruction better than the one it will supplant? Safety instruction attempts have had mixed reviews. Some studies report success in thwarting accidental injuries while others indicate failure and recommend that the instruction should be either revised or abandoned.

Advising. The effectiveness of this approach may be dependent upon the prestige of the person giving the advice. The prestige is related to the person's presumed experience, knowledge, and judgment. Examples of advising include: tornado and hurricane alerts, a state governor advising holiday travelers to obey speed limits, and traffic alerts during busy times alerting drivers to heavy traffic and traffic accidents.

Commanding. The adherence to a command also depends upon whether the commander has authority. Examples include: a policeman directing you to drive through a red light, a lifeguard whistling a person to stop running on the swimming pool deck, and a speed limit sign showing 55 mph.

Appealing to values. This appeal focuses upon the saving of a life or avoiding an injury. Examples of such appeals include: posters or signs showing and saying something about accident prevention and slogans such as "the life you save may be your own."

Inducements. This is an offering of something valued in return for safe behavior. Examples in industry are gifts (e.g., pens, trophies, camping equipment, etc.) given at the end of an accident-free period of time. Sometimes pictures and names appearing in a newspaper for safe behavior serve as an inducement.

Coercion. This is the opposite of inducement and means the threat of harm to the person. Examples of coercion include any threat by a parent to a child, whether it be a spanking or detainment in the child's room for the weekend if the child is misbehaving so as to risk the chance of an injury. Warning tickets given by a policeman serve as a threat to violators.

Legal sanctions. This is using force to make a person safe. This means changing behavior by requiring or prohibiting a certain type of behavior, by law or regulation. These include traffic laws such as running red lights, speeding, or driving while intoxicated. These are more politically acceptable than laws directed at protective behavior, such as required safety belt use or mandatory motorcycle helmet use. Having a law or regulation is not an assurance that accidents will decrease. Experience with the drinking driver problem and violations of the 55 mph speed limit law show that problems exist. Legal sanctions are usually the last effort for changing behavior to reduce accidents and injuries.

Changing or Establishing an Attitude

Acquiring or modifying an attitude is *not* done, according to a great deal of evidence, by the sole use of persuasive communication (e.g., repetition of "drive cautiously" or "please be careful").

An effective method of influencing an attitude is called "human modeling." A person can observe and learn attitudes from many sorts of human models. In one's younger years, one or both parents serve as models and later other members of the family may play this role. Teachers may become models for behavior, but the varieties of human modeling do not stop at the home or school. Public figures, prominent sports people, famous scientists or artists may become models. It is not essential that people who function as human models be seen or known personally— they can be seen on television or in movies. In fact, they can even be read about in books. This latter fact serves to emphasize the enormous potential that literature has for the determination of attitudes and values.

The human model must, of course, be someone whom the person respects, or as some writers would put it, someone with whom he or she can identify. The model must be observed (or read about) performing the desired kind of behavior. Having seen the action, whatever it may be, the person also must see that such action leads to satisfaction or pleasure on the part of the model (e.g., a movie star smiling while fastening his or her safety belt).

The modification of attitudes undoubtedly takes place all the time in every portion of an individual's life.

HADDON MATRIX

William Haddon, Jr.[2] developed a strategy system to aid in the identification of countermeasures. Though originally developed to cope with the traffic accident problem, the concept and matrix can be applied to any type of accident. The strategy is to reduce losses due to injuries rather than merely to prevent injuries. Haddon states that even when an accident cannot be prevented, there are many

Haddon suggests approaching the problem of reducing injuries by considering the three major phases that determine the final outcomes. These three phases are shown in Table 8-1 with examples of countermeasures related to crashes, burns, electrocutions, poisonings, and drownings.

The *first phase*, or pre-event phase, consists of many factors which determine whether an accident will take place. Elements that cause people and physical and/or chemical forces to move into undesirable interaction are included here. For example, probably the most important human factor in the pre-event phase is alcohol intoxication in almost all accident types.

In the past, the emphasis in accident prevention has been on human behavior

TABLE 8-1 Examples of Tactics for Reducing Injury Losses

TYPE OF EVENT	PRE-EVENT PHASE	EVENT PHASE	POST-EVENT PHASE
Impacts (e.g., from falls)	Alcoholism programs Handrails on stairs	Fire nets Padding on floors Football helmets	Trained ambulance crews Well-equipped ambulances Pneumatic splints
Exposure to heat	Child-proof matches Eliminating floor heaters Venting explosive gases	Flame-retardant clothing Reducing surface temperature of heaters and stoves Sprinkler systems in buildings	Burn centers Skin grafting Rehabilitation
Exposure to electricity	Covered electric outlets Insulation on electric handtools and wiring	Circuit breakers Fuses	Cardiopulmonary resuscitation Equipment and training
Ingestion of poison	Child-proof medicine containers Separation of CO from passenger compart- ments of autos	Making cleaning agents inert or less caustic Packing poisons in small, nonlethal amounts	Poison information centers Detoxification centers
Immersion in water	Fences around swimming pools Draining ponds	Life jackets Training to tread water or swim	Lifesaving training Teaching mouth-to-mouth re- suscitation techniques to general population

Source: Susan P. Baker and William Haddon, Jr., "Reducing Injuries and Their Results: The Scientific Approach," *The Milbank Memorial Fund Quarterly/Health and Society,* Fall 1974.

and attempts to change it. Accidents are usually regarded as someone's fault, rather than as a failure that could have been prevented by some change in the system. For example, if a boy, while cutting the grass with a power rotary lawn mower, ejects from the mower a stone which cuts a bystander, the resulting injury (the cut from the stone) is likely to be blamed on the boy operating the mower, rather than being attributed to the fact that a rotary mower many times will have no protective device to avert the ejection of stones and debris. Such devices are now required, but are often taken off.

Thus, the mower can contribute to the initiation of the accident, either by placing excessive demands or restrictions on the operator, or through mechanical inadequacies or failure. For example, in automobile crashes, steering, tire, and brake failures sometimes initiate crashes but seldom are searched for after the crash.

Other pre-event countermeasures relate to the environment, and there the principle of separation plays an important role. For example, children can be separated from cleaning agents containing caustic ingredients through the use of child-resistant containers and by storing containers in locked compartments out of reach of children.

The *second phase*, or event phase, begins when physical and/or chemical forces exert themselves unfavorably upon people and/or property. Countermeasures preventing harmful effects even when excessive energy (mechanical, chemical, thermal, etc.) is contacted are part of this second phase. Examples of event phase countermeasures for motor vehicle accidents include "packaging people" for crashes through the use of safety belts, padded dashes, and collapsible steering wheel columns. Nontraffic examples are boxing gloves, safety shoes, hard hats, helmets, lead x-ray shields, nets for acrobats, and gloves for laborers.

This second phase is to soften contact regardless of the cause. Stressing this phase follows the idea that because accidents will happen, let's protect humans and property the best we can. Tradition, on the other hand, has stressed the pre-event phase rather than the event phase.

The *third phase*, or post-event phase, involves salvaging people and/or property after contact with excessive energy (mechanical, chemical, thermal, etc.) has taken place. Early detection of aircraft crashes through transducers that start broadcasting a special signal at the time of a crash is an example of the third phase, as are fire detection systems (heat or smoke), SOS and MAYDAY signals, and the use of forest fire lookout towers.

In case of serious injury it is important to provide expert medical care as quickly as possible. Transportation for the injured, trained ambulance personnel, and emergency room staffs are appropriate countermeasures which are a part of this phase.

Haddon's other matrix consists of putting the three phases (pre-event, event, post-event) and the three epidemiological factors (human (host), agent, environment) together in a matrix which provides a greater practical and theoretical utility in categorizing countermeasure options. See Tables 8-2, 8-3, and 8-4.

It should be clear, however, that both of these matrices still encompass accident prevention efforts. The tables give examples of countermeasures for motor vehicle accidents, drowning, and suffocation in abandoned refrigerators. Not all of the possible countermeasures are given in the tables. Those given are for examples.

TABLE 8-2 Countermeasures. Accident Type: Motor Vehicle.

Phases / Factors	Pre-event	Event	Post Event
Host	Driver education	Driver "packaging"	Proper first aid and emergency care of bleeding
Agent	Automobile safety inspection of brakes, tires, etc.	Collapsible steering wheel to avoid impaling or crushing driver's chest	Accessible and low cost of vehicle-damage repair
Environment	Adequate signs and signals	Breakaway posts and sign poles	Emergency telephones and adequate emergency systems

Adapted from William Haddon, Jr., "A Logical Framework for Categorizing Highway Safety Phenomena and Activity," *Journal of Trauma*, March 1972, by permission of the author.

TABLE 8-3 Countermeasures. Accident Type: Drownings.

Phases / Factors	Pre-event	Event	Post Event
Host	Swimming instruction	Life jackets	Visible swimwear
Agent	No swimming pool	Shallow pool	Underwater lights
Environment	Barriers and fences	Lifelines	Rescue systems

Adapted from Park E. Dietz and Susan P. Baker, "Drowning Epidemiology and Prevention," *American Journal of Public Health*, April 1974.

THE TEN STRATEGIES

There are various additional ways to sort out options and tactics for reducing human and other damage. Haddon identified ten logically based strategies that are available and a guideline in formulating countermeasures.[3] All the strategies may be used for reducing the damage from all types of accidents.

These ten basic strategies, each with illustrative tactics, are:

1. To prevent the creation of the hazard in the first place. *Examples:* prevent production of boats, guns, snowmobiles.
2. To reduce the amount of the hazard brought into being. *Examples:* reduce speed of vehicles, lead content of paint, make less beverage alcohol (a hazard itself and in its results, such as drunken driving).
3. To prevent the release of the hazard that already exists. *Examples:* bolting or timbering mine roofs, impounding dangerous toys.
4. To modify the rate of spatial distribution of release of the hazard from its source. *Examples:* brakes, shutoff valves.

TABLE 8-4   Countermeasures. Accident Type:
Discarded or Abandoned Refrigerator Suffocation

Factors \ Phases	Pre-event	Event	Post Event
Host (human)	Tell children to stay away from discarded refrigerators because they are not "playthings" and they can kill	If a child is missing, a first place to look is in refrigerators	Resuscitation training for parents and older children
Agent	Manufacturers install permanent trays in refrigerators which don't allow room for children to crawl into them	Manufacturers install an interior wall section which a child's force can puncture to allow ventilation	Escape worthiness, (i.e., door can be opened from within by a force of 5 pounds)
Environment	Imposed penalties for discarding refrigerators without removing hinges and door	Manufacturers install an alerting device which would indicate occupancy and use (light or buzzer)	A first place to look for missing children is in refrigerators

5. To separate, in time or space, the hazard and that which is to be protected. *Examples:* walkways over or around hazards; evacuation; the phasing of pedestrian and vehicular traffic, whether in a work area or on a city street; the banning of vehicles carrying explosives from areas where they and their cargoes are not needed.

6. To separate the hazard and that which is to be protected by interposition of a material barrier. *Examples:* gloves, containment structures, child-proof poison-container closures, vehicle air bags.

7. To modify relevant basic qualities of the hazard. *Examples:* using breakaway roadside poles, making crib slat spacing too narrow to strangle a child.

8. To make what is to be protected more resistant to damage from the hazard. *Examples:* making structures more fire- and earthquake-resistant, giving salt to workers under thermal stress, making motor vehicles more crash resistant.

9. To begin to counter the damage already done by the environmental hazard. *Examples:* rescuing the shipwrecked, reattaching severed limbs, extricating trapped miners.

10. To stabilize, repair, and rehabilitate the object of the damage. *Examples:* posttraumatic cosmetic surgery, physical rehabilitation for amputees and others with disabling injuries (including many thousands paralyzed annually by spinal cord damage sustained in motor-vehicle crashes), rebuilding after fires and earthquakes.

Two points should be kept in mind in using these strategies: (1) they provide guidelines for possible control programs; they do not provide a formula or guide for specific cases. These should be dealt with on an individual basis; and (2) the strategies do not center on causation but instead on the entire realm of how to reduce damages.

Ten Strategies Example

Because of their importance, sports injuries will be used to illustrate approaches based on *Haddon's ten basic strategies for preventing injury.*[4]

The *first strategy is to prevent the creation of the hazard in the first place*, for example, by not manufacturing sports equipment that is apt to cause injury. This strategy might be applied to trampolines, an important source of spinal cord injuries. After the American Academy of Pediatrics recommended in 1977 that school use of trampolines be banned, there was a drop of more than 60 percent in the number of trampoline head and neck injuries treated in hospitals participating in the Consumer Product Safety Commission's National Electronic Injury Surveillance System (NEISS).

The *second strategy is to reduce the amount of hazard that is created*, for example, reducing the height from which people can fall or jump, limiting the speed capability of snowmobiles, or limiting the speed of beginning skiers by providing slopes with only small vertical drops in relation to the lengths of the trails. Exposure can be curtailed through shorter periods of play or by permitting hunting only on certain days. Reducing the number of players who participate in a particular sport is another example, illustrated by limiting participants to members of a specified age group.

The *third strategy involves either preventing or reducing the likelihood of the release of a hazard.* Examples include not allowing boxers to fight and designing hunting weapons that will not discharge inadvertently. In ancient times, the cessation of gladiatorial contests provides an additional example, as would the ending of bull fighting in Latin countries. Reducing the likelihood of the release of the hazard is often a more practical approach; an example is packing and grooming ski slopes to reduce hidden obstacles that might cause skiers to fall.

The *fourth strategy is to modify the rate or spatial distribution of release of the hazard from its source.* Examples include release bindings on skis, controlled release of dammed-up water to protect boaters downstream, and the use of shorter cleats on football shoes so that the foot can rotate easily without transmitting a sudden force to the knee. Changes in football rules have outlawed spearing and face-tackling; these techniques use the head as a primary contact point, and the forces on the head and neck are likely to exceed injury thresholds. Yet, more than one-third of high school football players in Minnesota continued to use these maneuvers a year after they were banned, and one player in five reported concussion symptoms during the playing season.

The *fifth strategy is to separate, in time or space, the hazard and its release from that which is to be protected.* Starting avalanches during times when ski slopes are closed, an example of temporal separation, decreases the likelihood that avalanches will occur when skiers are on the slopes. Placing benches and other equipment farther from playing areas reduces the frequency of "out of bounds" injuries that commonly occur when players run into them. Spatial separation is also illustrated by storing pistols used for target shooting at the shooting range rather than at home, and by providing paths for bicyclists, joggers, and people walking that are separate from roads for motor vehicles.

The *sixth strategy is to separate the hazard from whatever is to be protected by interposing a material barrier.* In many sports, the head, face, eyes, chest, or other body parts need to be protected from balls, bats, or other players. A review of sports-related injuries and deaths among people ages five to fourteen revealed that 38 percent of the deaths involved baseball. Being struck on the chest by the baseball with subsequent cardiac arrest appeared to be the predominant cause. This suggests a need for chest protection for young baseball players. Eye protection devices for racquetball and squash can prevent many eye injuries, the most common serious injury associated with such racquet sports. Facial and dental injuries among hockey and football players have been substantially reduced by face masks. Protective helmets are appropriate to many sports, such as football and horseback riding, where head injuries are a serious problem.

The *seventh strategy is to modify the relevant basic qualities of the hazard.* Illustrations include recent adoption of a softer ball in squash rather than the previously used hardball, padding the outer edge of racquets and using balls large enough so that the bony socket of the eye affords some protection. The pointed ends of hockey sticks, once a major source of facial injuries, are now rounded to make them less injurious. Gymnasium walls should be designed without protrusions and either made of energy-attenuating materials or padded in areas where players can strike them. Breakaway goalposts and slalom poles that yield on impact are further examples.

The *eighth strategy is to make that which is to be protected more resistant to damage from the hazard.* Conditioning of the musculo-skeletal system is an important means of reducing the likelihood of injury. Grouping school athletes by skills, physical fitness, and physical maturity rather than age has reportedly reduced injury rates. Exercise and therapy to reduce osteoporosis are promising approaches of special relevance to older people participating in athletic and recreational activities.

The *ninth strategy is to begin to counter damage already done.* Athletes who may have sustained spinal cord injuries, for example, need to be carefully supported when they are moved in order to reduce the likelihood of paralysis. Football players with concussion symptoms should not be returned to play on the same day because of the potential for progressive neurological debilitation. One study found that most high school players who experienced loss of consciousness returned to play the same day. (Communication systems and readily available emergency and definitive care are clearly important but often inadequate, except in some major urban centers and in states with good emergency systems.

The *tenth strategy is to stabilize, repair, and rehabilitate the injured person.* Reconstructive surgery, physical and mental rehabilitation, and modification of the environment to accommodate the handicapped help to minimize adverse outcomes of serious injury.

These ten strategies and examples of illustrative tactics suggest the wide variety of measures that can reduce the likelihood and severity of injuries, as well as the severity of the consequences of injury once it has occurred. In choosing among potentially useful preventive measures, priority should be given to the ones most likely to effectively reduce injuries. In general, these will be measures that provide built-in automatic protection, minimizing the amount and frequency of effort required of the individuals involved.

NOTES

1. Author unknown.
2. William Haddon, Jr., "A Logical Framework for Categorizing Highway Safety Phenomena and Activity," *The Journal of Trauma*, 12 (1972), 193–207.
3. William Haddon, Jr., "Advances in the Epidemiology of Injuries As a Basis for Public Policy," *Public Health Reports*, 95 (1980), 411–21, and William Haddon, Jr., "The Basic Strategies for Reducing Damage from Hazards of All Kinds," *Hazard Prevention*, 16 (1980), 1.
4. Susan P. Baker et al., *The Injury Fact Book* (Lexington, MA: Lexington Books, 1984), pp. 94–97.

Index